Coding Theory Essentials

Edited by Dinesh G. Harkut
and Kashmira N. Kasat

Published in London, United Kingdom

Coding Theory Essentials
http://dx.doi.org/10.5772/intechopen.104071
Edited by Dinesh G. Harkut and Kashmira N. Kasat

Contributors
Boris Assanovich, Iryna Korlyukova, Andrei Khombak, Rita Vincenti, Kais Hassan, Kosai Raoof, Pascal Chargé, Asma'a Al-Hakimi, Abu Bakar Md Sultan, Chen Chen, Fengfan Yang, Venkata Naga Srinivasa Rao Repalle, Keneni Abera Tola, Mamo Abebe Ashebo, Manjeet Singh Teeth, Sanjay Harne, Dinesh G. Harkut, Kashmira N. Kasat

First published in London, United Kingdom, 2023 by IntechOpen
IntechOpen is the global imprint of INTECHOPEN LIMITED, registered in England and Wales, registration number: 11086078, 5 Princes Gate Court, London, SW7 2QJ, United Kingdom

British Library Cataloguing-in-Publication Data
A catalogue record for this book is available from the British Library

Additional hard and PDF copies can be obtained from orders@intechopen.com

Coding Theory Essentials
Edited by Dinesh G. Harkut and Kashmira N. Kasat
p. cm.
Print ISBN 978-1-83768-502-8
Online ISBN 978-1-83768-503-5
eBook (PDF) ISBN 978-1-83768-504-2

Meet the editor

Dr. Dinesh G. Harkut is an associate professor and head of the Computers Science and Engineering Department, Prof Ram Meghe College of Engineering & Management (PRMCEAM), Badnera, India. He is a pioneering researcher in soft computing and embedded systems with dual doctorates in Computer Science and Engineering and Business Management. His primary research interests include computer artificial intelligence (AI), big data, analytics, embedded systems, and e-commerce. He has supervised eighteen master's degree and forty bachelor's degree students. He has published forty-eight papers in refereed journals and published eight books with international publishers. He holds four registered patents. Dr. Harkut is the principal investigator in setting centers of excellence for renowned technological giants like IBM, Oracle, Texas Instruments, and Huawei at PRMCEAM and establishing industry-funded laboratories for ARM, Cypress Semiconductor, Intel FPGA, Wind River, and Xilinx. He is a fellow member of the Institute of Electronics and Telecommunication Engineering (IETE), New Delhi; a life member of the Indian Society for Technical Education (ISTE), New Delhi, and the Computer Society of India (CSI); senior member of the Universal Association of Computer and Electronics Engineers (UACEE), USA; and professional member of the International Association of Engineers (IAENG), Hong Kong.

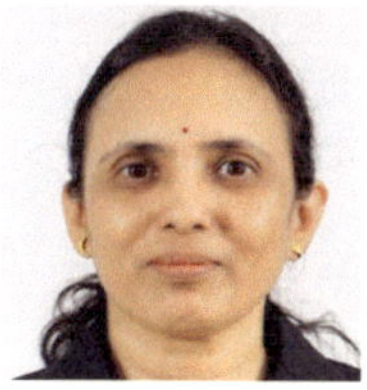

Dr. Kashmira N. Kasat received her Ph.D. in Electronics Engineering from Sant Gadge Baba Amravati University (SGBAU), Amravati, Maharashtra, India, in 2017. She obtained her Graduation B.E. (Ind. Electronics) from Jawaharlal Nehru Engineering College (JNEC), Aurangabad, India in 1999 and a Post-graduation M.E. (Electronics) from Government College of Engineering (GCOE), Aurangabad, India, in 2009. Presently, she is an assistant professor and head of the Electronics and Telecommunication Department at Prof Ram Meghe College of Engineering & Management (PRMCEAM). She has 12 years of teaching and industrial experience. She has published seventeen journal papers and one book. She also holds three patents. Dr. Kasat has supervised eighteen undergraduate and seven postgraduate projects. Her areas of research include power electronics, soft computing, and very large-scale integration VLSI design. She is a life member of the International Society for Technology in Education (ISTE) and Institution of Electronics and Telecommunication Engineers (IETE) and a senior member of the Universal Association of Computer and Electronics Engineers (UACEE). She is an editorial board member of the Scientific Board of Computer, Electrical & Electronic Engineers, International Institute of Engineers, UK.

Contents

Preface

"Code is poetry, and coding theory is the symphony that orchestrates the harmony of information"

Coding Theory Essentials is an edited book that unveils the intricacies and wonders of a field that lies at the heart of modern communication and information storage. In this digital era, where the transfer and protection of data have become paramount, coding theory plays a vital role in ensuring the reliability and security of our interconnected world.

The essence of coding theory is the art of transforming messages into a language that transcends the boundaries of noise, interference, and errors. By cleverly encoding information and implementing efficient algorithms, coding theorists have paved the way for robust and error-resilient systems, enabling us to transmit data over vast distances and safeguard it from corruption.

This book serves as a comprehensive guide, capturing the fundamental concepts, principles, and techniques that underpin coding theory. Whether you are an aspiring computer scientist, an electrical engineer, or a curious mind seeking to delve into the captivating realm of information theory, this compilation of knowledge will empower you to understand, appreciate, and apply the core tenets of coding theory.

The chapters within have been meticulously crafted by experts in the field, combining theoretical foundations with real-world applications. Through a step-by-step journey, we explore diverse topics, ranging from error-correcting codes and block codes to convolutional codes, channel capacity, and beyond. Each chapter offers a careful balance between theory and practical insights, equipping readers with both the conceptual understanding and the tools necessary to address modern challenges in data transmission and storage.

As you embark on this intellectual expedition, we encourage you to embrace the beauty and elegance that lies within coding theory. Witness how seemingly abstract mathematical concepts transform into tangible solutions that revolutionize our daily lives. Discover the profound impact of error-correcting codes on digital communication, from wireless networks to satellite transmissions, from data centers to deep-space exploration.

Our aim in creating this book is to inspire and empower readers to explore, experiment, and innovate within the realm of coding theory. We hope that the insights gained from these pages will spark new ideas, foster interdisciplinary collaborations, and ignite a passion for the intricate world of information coding.

We extend our deepest gratitude to the contributing authors who have dedicated their expertise, time, and energy to enriching this collection. Their collective wisdom and dedication have made this book a valuable resource for researchers, students, and practitioners alike.

Finally, we would like to express our gratitude to the readers, whose curiosity and thirst for knowledge motivate us to bring this book to fruition. May *Coding Theory Essentials* serve as a guiding light on your path towards unraveling the mysteries of coding theory and embracing the boundless possibilities it offers.

I owe special thanks to IntechOpen for their kind support and great efforts in bringing the book to fruition. I also appreciate all those who worked tirelessly in the background throughout the publication process.

Dr. Dinesh G. Harkut
Head and Associate Professor,
Department of Computer Science and Engineering,
Professor Ram College of Engineering and Management,
Amravati, M.S., India

Dr. Kashmira N. Kasat
Head and Assistant Professor,
Department of Electronics and Telecommunication Engineering,
Professor Ram College of Engineering and Management,
Amravati, M.S., India

Chapter 1

Introductory Chapter: Coding Theory

Dinesh G. Harkut and Kashmira N. Kasat

1. Introduction

The field of channel coding is a fundamental part of digital communication systems. Its purpose is to enable reliable transmission of data over noisy and error-prone communication channels. Channel coding theory deals with the design of error-correcting codes that can tolerate a certain number of errors introduced during transmission, while still allowing for accurate reconstruction of the original data at the receiver end.

Channel coding is the process of adding redundancy to a message or data stream to protect against errors that may occur during transmission over a noisy communication channel. The redundant information, known as the error-correcting code, enables the receiver to detect and correct errors that occur during transmission. For example, let us say you want to send a message "HELLO" to your friend over a communication channel. During transmission, the message may get corrupted due to noise or interference on the channel. To protect against errors, you can add redundancy to the message by encoding it with an error-correcting code, such as a cyclic redundancy check (CRC) code. The CRC code adds a checksum to the message, which is computed based on the message content using a mathematical algorithm. The receiver also computes the checksum of the received message and compares it with the checksum sent by the transmitter. If the two checksums do not match, it indicates that there was an error in the transmission, and the receiver requests the transmitter to resend the message. In this example, the CRC code serves as a channel coding scheme that protects the message against errors during transmission over the noisy communication channel. By adding redundancy to the message, the receiver can detect and correct errors, ensuring reliable communication between the transmitter and receiver.

2. Historical background

The history of channel coding dates back to the early days of telegraphy and radio communication, when engineers first realized the need for error-correcting codes to ensure reliable transmission over noisy and error-prone communication channels. The first error-correcting codes were developed in the 1940s, with the work of Richard Hamming and Claude Shannon laying the foundation for modern channel coding theory [1]. Richard Hamming developed the concept of Hamming codes in 1950, which were the first practical error-correcting codes used in digital communication systems. These codes were designed to detect and correct single-bit errors, and they were widely used in early computer systems and communication protocols [1]. Claude

Shannon, a pioneer in information theory, introduced the concept of channel capacity in his landmark paper "A Mathematical Theory of Communication" published in 1948 [1]. Shannon's work established the fundamental limits of communication over noisy channels, and it paved the way for the development of more sophisticated error-correcting codes that could approach these limits. Since then, channel coding theory has continued to evolve, with new codes and techniques being developed to address the challenges of modern communication systems. Today, channel coding is an essential part of digital communication systems, enabling reliable transmission of data over a wide range of communication channels.

3. Basic concepts

Channel coding theory is based on several fundamental concepts that are essential to understanding how error-correcting codes work. These concepts include code rate, block length, minimum distance, error correction capability, and channel capacity.

- Code rate: The code rate is the ratio of the number of message bits to the total number of transmitted bits, including the redundant bits added by the error-correcting code. A higher code rate means that more information is transmitted per bit, but it also means that the error-correction capability of the code is reduced [2].

- Block length: The block length is the number of bits in a block of data that is encoded using an error-correcting code. Longer block lengths typically provide better error-correction capabilities, but they also increase the delay and complexity of the encoding and decoding processes [2].

- Minimum distance: The minimum distance of an error-correcting code is the smallest number of bit changes that must occur to transform one valid codeword into another. A higher minimum distance means that the code is more robust against errors, as it can detect and correct more errors during transmission [2].

- Error correction capability: The error correction capability of an error-correcting code is the maximum number of errors that can be corrected during transmission. This capability depends on the code rate, block length, and minimum distance of the code, as well as the characteristics of the communication channel [2].

- Channel capacity: The channel capacity is the maximum rate at which information can be transmitted over a noisy communication channel with a given error rate. This limit is determined by the channel characteristics and the laws of physics, and it provides a theoretical upper bound on the performance of any error-correcting code used over the channel [2].

These basic concepts form the foundation of channel coding theory, and they are used to design and analyze error-correcting codes for a wide range of communication systems. By understanding these concepts, engineers can develop more efficient and effective error-correcting codes that can provide reliable communication over noisy and error-prone channels. Several types of error-correcting codes are commonly used in digital communication systems. Each type has its own advantages and disadvantages, depending on the specific application requirements and constraints. The main types of error-correcting codes are:

4. Block codes

Block codes divide a message into fixed-size blocks, and each block is encoded separately using an error-correcting code. The most common block codes are Reed-Solomon codes, which are used in a wide range of applications, including CD and DVD storage, satellite communication, and digital television [3].

Advantages:

- High error-correction capability
- Simple encoding and decoding algorithms
- Robust against burst errors

Disadvantages:

- High redundancy, leading to lower code rate
- Inefficient for variable-length messages

5. Convolutional codes

Convolutional codes are based on a mathematical concept called convolution, which involves multiplying and adding a sequence of numbers. Convolutional codes are designed to operate on a continuous stream of data, and they use a sliding window to encode and decode the data [2].

Advantages:

- High error-correction capability
- Efficient for variable-length messages
- Can be used for high-speed communication systems

Disadvantages:

- Complex encoding and decoding algorithms
- Sensitive to phase distortion and timing errors
- Limited ability to correct burst errors

6. Turbo codes

Turbo codes are a type of iterative code that use multiple convolutional codes in parallel, with a feedback mechanism to refine the decoding process. Turbo codes are widely used in mobile communication systems, such as 3G and 4G cellular networks [4].

Advantages:

- Very high error-correction capability
- Efficient for variable-length messages
- Robust against noise and interference

Disadvantages:

- High complexity, requiring specialized hardware or software
- Sensitive to phase distortion and timing errors
- Limited availability of standards and implementation tools

7. LDPC codes

Low-density parity-check (LDPC) codes are a type of linear block code that use a sparse parity-check matrix to encode and decode data. LDPC codes are widely used in high-speed communication systems, such as wired and wireless networks, as well as storage systems [5].

Advantages:

- High error-correction capability
- Efficient for variable-length messages
- Robust against noise and interference

Disadvantages:

- Complex encoding and decoding algorithms
- Sensitive to channel characteristics and noise models
- Limited availability of standards and implementation tools

Overall, the choice of error-correcting code depends on the specific application requirements and constraints, such as the required error-correction capability, message length, transmission rate, and implementation complexity. Each type of error-correcting code has its own trade-offs between error-correction capability, efficiency, complexity, and robustness, and the most appropriate code must be selected based on a careful analysis of the application requirements and constraints.

8. Encoding and decoding

Encoding and decoding are fundamental operations in channel coding theory, which are used to convert an input message into a coded message and to recover the original message from the received coded message, respectively. The encoding and decoding processes are based on mathematical algorithms and principles, which are designed to provide a certain level of error-correction capability and robustness to the transmission of data over noisy and unreliable communication channels [3].

Encoding: Encoding involves transforming an input message into a coded message, which contains additional redundancy information that can be used to detect and correct errors that occur during transmission. The encoding process typically involves applying a mathematical function or algorithm to the input message, which generates a set of parity bits that are added to the message to form the coded message. For example, consider a simple block code called a parity code, which involves adding a single parity bit to a message of length n bits. The parity bit is computed as the XOR (exclusive OR) of all the bits in the message, and it is added to the end of the message to form the coded message. The encoding process can be represented by the following equation:

$$C = M \| P \tag{1}$$

where M is the original message, P is the parity bit, || denotes concatenation, and C is the coded message. For example, if the input message is 1011, the parity bit is computed as 1 XOR 0 XOR 1 XOR 1 = 1, and the coded message is 10111.

Decoding: Decoding involves recovering the original message from the received coded message, which may have been corrupted by errors during transmission. The decoding process typically involves applying a mathematical function or algorithm to the received coded message, which uses the redundant information in the message to detect and correct errors and recover the original message. For example, consider the parity code described above. To decode a received message, the receiver computes the parity bit of the received message and compares it to the received parity bit. If they are the same, the message is assumed to be error-free and the original message is recovered by removing the parity bit. If they are different, an error is detected and the receiver may attempt to correct the error by flipping the received bit that is inconsistent with the computed parity bit. The decoding process can be represented by the following equation:

$$M' = C[1 : n]. \tag{2}$$

where C is the received coded message, M′ is the recovered message, and [1:n] denotes the first n bits of C. Of course, this is just a simple example of encoding and decoding with a parity code, and there are many more sophisticated and powerful coding schemes that are used in practice. However, the basic principles of encoding and decoding remain the same, and they are critical for ensuring reliable and efficient communication over noisy and unreliable channels.

There are various implementation strategies available for channel coding, and the choice of implementation strategy depends on factors such as the complexity of the coding scheme, the required data rate, and the hardware and software resources available for implementation. In this section, we will discuss some common implementation strategies for channel coding, along with their advantages and disadvantages and examples of their applications in real-time environments.

Software-based implementation: Software-based implementation of channel coding involves implementing the encoding and decoding algorithms using software running on a general-purpose processor such as a CPU or a DSP. This implementation strategy is flexible and can be easily updated or modified as needed, but may be slower and less power-efficient compared to hardware-based implementation.

Example: The Reed-Solomon code is a widely used block code that can correct multiple errors in a block of data. Reed-Solomon coding is often implemented in software on general-purpose processors for applications such as digital audio and video storage and transmission, and satellite communication.

FPGA-based implementation: FPGA (Field-Programmable Gate Array) based implementation of channel coding involves implementing the encoding and decoding algorithms on an FPGA, which is a programmable hardware device that can be reconfigured to perform different functions. This implementation strategy provides high performance and low latency, but may require specialized expertise and tools for design and implementation.

Example: The Turbo code is a powerful and widely used convolutional code that can achieve very high data rates and error-correction capability. Turbo code decoding is often implemented on FPGAs for applications such as wireless communication, digital broadcasting, and satellite communication.

ASIC-based implementation: ASIC (Application-Specific Integrated Circuit) based implementation of channel coding involves designing and fabricating custom hardware circuits that implement the encoding and decoding algorithms. This implementation strategy provides high performance and low power consumption, but may require high initial costs and long design and fabrication times.

Example: The LDPC (Low-Density Parity-Check) code is a powerful and efficient linear code that can achieve very high data rates and error-correction capability. LDPC code decoding is often implemented on ASICs for applications such as wireless communication, digital broadcasting, and storage systems.

Hybrid implementation: Hybrid implementation of channel coding involves combining different implementation strategies such as software, FPGA, and ASIC to achieve the desired balance of performance, flexibility, and cost. This implementation strategy can provide high performance and flexibility while reducing the costs and development time compared to fully custom hardware implementation.

Example: The convolutional code is a popular and widely used linear code that can achieve high data rates and error-correction capability. Convolutional code decoding is often implemented using a hybrid implementation strategy that combines software and FPGA or ASIC for applications such as wireless communication and digital broadcasting.

The choice of implementation strategy for channel coding depends on the specific requirements and constraints of the application. Software-based implementation provides flexibility and ease of development, FPGA-based implementation provides high performance and low latency, ASIC-based implementation provides high performance and low power consumption, and hybrid implementation provides a balance of performance and cost.

Channel coding theory is a fundamental aspect of modern communication systems that enables reliable transmission of digital data over noisy communication channels. The goal of channel coding is to add redundancy to the transmitted data in such a way that the receiver can correct errors caused by noise or interference in the channel. This is achieved by using error-correcting codes that add redundancy to the data stream, which can be used by the receiver to detect and correct errors. Channel coding theory involves the study of the mathematical principles underlying error-correcting codes, their encoding and decoding algorithms, and their performance analysis. Channel coding is a multidisciplinary field that draws upon concepts from mathematics, computer science, information theory, and communication engineering.

The most common types of error-correcting codes used in channel coding are block codes and convolutional codes. Block codes divide the input data into fixed-size blocks and add parity bits to each block, while convolutional codes operate on a continuous stream of input data and add redundant symbols based on a sliding window of previous symbols. One important concept in channel coding theory is the Hamming distance, which is a measure of the number of bit positions in which two binary strings differ. The minimum Hamming distance of an error-correcting code is the smallest Hamming distance between any two valid codewords, and it determines the error-correction capability of the code.

Another important concept in channel coding theory is the decoding algorithm, which is used by the receiver to recover the original data from the received codeword. There are two main types of decoding algorithms: maximum likelihood decoding and syndrome decoding. Maximum likelihood decoding involves searching for the most likely codeword given the received data, while syndrome decoding involves using the syndrome of the received codeword to correct errors. Channel coding theory also includes the analysis of the performance of error-correcting codes under different channel conditions, such as the signal-to-noise ratio (SNR) and the bit error rate (BER). The performance of a code is typically measured by its error-correction capability, which is the maximum number of errors that the code can correct, and its coding efficiency, which is the ratio of the number of information bits to the total number of transmitted bits.

The development of channel coding theory has led to the discovery of many powerful error-correcting codes that have found widespread use in communication systems. Some examples of widely used codes include the Reed-Solomon code, which is used in digital audio and video storage and transmission, the Turbo code, which is used in wireless communication and digital broadcasting, and the LDPC code, which is used in wireless communication and storage systems.

In conclusion, channel coding theory is a crucial aspect of modern communication systems that enables reliable transmission of digital data over noisy channels. It involves the study of error-correcting codes, their encoding and decoding algorithms, and their performance analysis. The development of channel coding theory has led to the discovery of many powerful error-correcting codes that have found widespread use in communication systems.

Author details

Dinesh G. Harkut[1*] and Kashmira N. Kasat[2]

1 Department of Computer Science and Engineering, Prof. Ram Meghe College of Engineering and Management, Badnera-Amravati, M.S., India

2 Department of Electronics and Telecommunication Engineering, Prof. Ram Meghe College of Engineering and Management, Badnera-Amravati, M.S., India

*Address all correspondence to: dg.harkut@gmail.com

References

[1] Shannon CE. A mathematical theory of communication. The Bell System Technical Journal. 1948;**27**(3):379-423

[2] Proakis JG, Salehi M. Digital Communications. 5th ed. New York, NY, USA: McGraw-Hill; 2008

[3] Lin C, Costello DJ Jr. Error Control Coding: Fundamentals and Applications. Upper Saddle River, NJ, USA: Pearson/ Prentice Hall; 2004

[4] Berrou C, Glavieux A, Thitimajshima P. Near Shannon limit error-correcting coding and decoding: Turbo-codes. In: Proceedings of the International Conference on Communications (ICC'93); Geneva, Switzerland. 1993. pp. 1064-1070

[5] Gallager RG. Low-density parity-check codes. IRE Transactions on Information Theory. 1962;**8**(1):21-28

Chapter 2

RS Codes and Optimized Distributed RS-Coded Relay Cooperative Communications: Code Constructions and Performance Analysis

Chen Chen and Fengfan Yang

Abstract

This chapter introduces the Reed-Solomon (RS) codes and the distributed RS-coded cooperative system over the Rayleigh fading channel, where the encoding and decoding procedures of the RS codes are elaborated. Besides, two optimized selection approaches, i.e., the exhaustive search approach and partial search approach, are employed in the relay to obtain a resultant code at the destination with better weight distribution. Moreover, the two joint decoding algorithms, namely naive and smart algorithms, are presented that further improve the overall average bit error rate (BER) performance of the cooperative scheme. Also, the performance analysis of the distributed RS-coded cooperative scheme is provided in detailed.

Keywords: BCH codes, RS codes, relay cooperation, distributed RS codes, joint decoding

1. Introduction

Fifth-generation (5G) communication systems may accommodate the traffic generated by a variety of wireless network types such as Device-to-Device (D2D) and sensor networks. Hence, it is reasonable to consider the short-information-transmission scenario. Generally, one of the most important aspects of transmission is to combat the signal fading over a wireless channel. Spatial diversity has proven to be the most effective method for mitigating the impacts of fading [1]. However, many mobile communication devices are unable to leverage spatial diversity techniques owing to size, power, and hardware complexity. Therefore, coded cooperative diversity with the aid of the relay was proposed to provide uplink diversity via single antenna sharing. Factually, various distributed linear block codes have been employed in the coded cooperation such as the distributed turbo codes (DTC) [2], distributed low-density parity-check codes (D-LDPC) [3], and polar codes [4]. Nevertheless, for the non-binary codes with short information sizes in coded cooperation, the literature has not been thoroughly investigated. Note that Reed-Solomon (RS) codes are a well-known class of non-binary codes with low encoding and decoding complexity. Furthermore, as a member of maximum distance separatable (MDS) codes, short-to-medium-length RS codes perform well in correcting random burst errors. Hence, RS-coded relay cooperation is considered a promising exploration to support short information transmission [5]. In addition, the distinct information selection in the relay may result in a different resultant code at the destination, which will influence the performance of the overall transmission. Hence, the optimized selection approaches [6] at the relay are also introduced in this chapter.

The remaining contexts of this chapter are summarized as follows. Section 2 provides a brief introduction to the BCH codes and RS codes. The general distributed RS-coded cooperative system is presented in Section 3. Section 4 exhibits the two optimized selection approaches and the corresponding examples. The joint decoding algorithms and the performance analysis are elaborated in Section 5. Section 6 concludes this chapter.

2. BCH codes and RS codes

2.1 BCH codes

Bose-Chaudhuri-Hocquenghem (BCH) codes are a kind of cyclic codes that can effectively correct random errors [7], which can be classified into binary BCH codes and non-binary BCH codes according to the different fields from which symbols are taken. Given any finite field GF(q) and its extension field $\mathrm{GF}(q^{\mathrm{m}})$, where q is a prime or a power of a prime and m is a positive integer, let α be a non-zero and non-one element of $\mathrm{GF}(q^{\mathrm{m}})$. If the generator polynomial $g(x) \in F[x](F \in GF(q))$ is the lowest-degree-polynomial with consecutive roots $\{\alpha, \alpha^2, \cdots, \alpha^{2t}\}$, then a cyclic code generated from this polynomial $g(x)$ is called a BCH code.

Assume that $\varphi_i(x)$ denotes the minimum polynomial of $\alpha^i (1 \le i \le 2t)$ and e^i represents the order of α^i. Therefore, the generator polynomial $g(x)$ and the code length n of BCH code are provided as,

$$g(x) = LCM\{\varphi_1(x), \varphi_2(x), \dots, \varphi_i(x)\}, n = LCM\{e_1, e_2, \dots, e_2t\}, \tag{1}$$

where LCM denotes the least common multiple. In particular, when $q = 2$, it is the binary BCH code. Also, if α is the primitive element in $\mathrm{GF}(q^{\mathrm{m}})$, it is a primitive BCH code of code length $n = q^m - 1$. Otherwise, the BCH code is non-primitive where n is the factor of $q^m - 1$. Consider a BCH code of length n, its parity check matrix is provided as [8],

$$H = \begin{bmatrix} 1 & \alpha & \alpha^2 & \alpha^3 & \cdots & \alpha^{n-1} \\ 1 & \alpha^2 & (\alpha^2)^2 & (\alpha^2)^3 & \cdots & (\alpha^2)^{n-1} \\ 1 & \alpha^3 & (\alpha^3)^2 & (\alpha^3)^3 & \cdots & (\alpha^3)^{n-1} \\ \vdots & \vdots & \vdots & \vdots & \ddots & \vdots \\ 1 & \alpha^{2t} & (\alpha^{2t})^2 & (\alpha^{2t})^3 & \cdots & (\alpha^{2t})^{n-1} \end{bmatrix}. \tag{2}$$

Then, the minimum distance of the t-error-correcting BCH codes is at least $2t+1$. The proof process can be referred to [9]. This lower bound on the minimum distance is called the BCH bound.

2.2 RS codes: encoding and decoding

The most important subclass of q-ary BCH codes is the RS codes, a particular subclass of q-ary BCH codes for which $m=1$. The efficient encoding and hard-decision decoding algorithms of RS codes as well as their improved capacity to rectify random burst errors have made them extensively applied for error control in both storage systems and digital communication [9]. The following describes the specific characteristic, encoding, and decoding processes of the RS codes.

2.2.1 Free distance of RS codes

Suppose that α is a primitive element in $GF(q)$. The generator polynomial g(x) of t-error-correcting (n,k) RS code has $\{\alpha, \alpha^2, \cdots, \alpha^{2t}\}$ as all its roots, where all symbols of RS codes are chosen from $GF(q)$, n and k denote the code length and length of information sequence, respectively. Therefore, the minimum polynomial $\varphi_i(x)$ corresponding to each α^i is $x-\alpha^i$. And $g(x)$ can be obtained from Eq. (1) given as,

$$\begin{aligned} g(x) &= (x-\alpha)(x-\alpha^2)\cdots(x-\alpha^{2t}) \\ &= g_0 + g_1x + g_2x^2 + \cdots + g_{2t-1}x^{2t-1} + x^{2t}, \end{aligned} \tag{3}$$

where $g_i \in GF(q)$ for $0<i<2t$. Since the all roots of $x^{q-1}-1$ are $\alpha, \alpha^2, \ldots, \alpha^{2t}, g(x)$ can divides $x^{q-1}-1$. Thus, $g(x)$ generates a q-ary RS code of length $n=q-1$ with exactly $2t$ parity-check symbols, which means $n-k=2t$.

From the BCH bound and the Eq. (3) where the code polynomial comprises $2t+1$ terms. Hence, there cannot be a zero for any of the coefficients in $g(x)$ can be zero. Otherwise, the resultant codeword would have a weight less than $2t+1$, which would be in conflict with the BCH bound on the minimum distance. As a result, the $g(x)$ corresponds to a codeword with a weight of precisely $2t+1$. It follows that the minimum distance of the t-error-correcting RS code generated by Eq. (3) is determined as exactly $2t+1$, i.e., $d_{min}=2t+1$. In addition, the minimum distance of the RS code is more than the number of its parity-check symbols. Therefore, RS codes are a prominent subgroup of the maximum distance separable (MDS) codes [10]. In this chapter, we simply consider $q=2$.

Example 1. Let α is a primitive element in $GF(2^4)$ constructed based on the primitive polynomial $1+x+x^4$ shown in **Table 1**. Consider the double-error-correcting RS codes with the symbols from $GF(2^4)$. The generator polynomial $g(x)$ of this code has $\alpha, \alpha^2, \alpha^3, \alpha^4$ as all its roots. Hence, $g(x)$ is acquired as,

$$\begin{aligned} g(x) &= (x-\alpha)(x-\alpha^2)(x-\alpha^3)(x-\alpha^4) \\ &= \alpha^{10} + \alpha^3x + \alpha^6x^2 + \alpha^{13}x^3 + x^4, \end{aligned} \tag{4}$$

The code is (15,11) RS code over the $GF(2^4)$ that can correct two errors. And, the minimum distance of this RS code is 5.

The end of Example 1.

Field elements	Vector	Field elementst	Vector
0	[0000]	$\alpha^7 = 1+\alpha+\alpha^3$	[1101]
	[1000]	$\alpha^8 = 1+\alpha^2$	[1010]
α	[0100]	$\alpha^9 = \alpha+\alpha^3$	[0101]
α^2	[0010]	$\alpha^{10} = 1+\alpha+\alpha^2$	[1110]
α^3	[0001]	$\alpha^{11} = \alpha+\alpha^2+\alpha^3$	[0111]
$\alpha^4 = 1+\alpha$	[1100]	$\alpha^{12} = 1+\alpha+\alpha^2+\alpha^3$	[1111]
$\alpha^5 = \alpha+\alpha^2$	[0110]	$\alpha^{13} = 1+\alpha^2+\alpha^3$	[1011]
$\alpha^6 = \alpha^2+\alpha^3$	[0011]	$\alpha^{14} = 1+\alpha^3$	[1001]

Table 1.
Galois field $GF(2^4)$ with the primitive polynomial $1+\alpha+\alpha^4 = 0$.

2.2.2 Encoding of RS codes

Given the generator polynomial $g(x)$ illustrated in Eq. (3), the polynomial $c(x)$ of the codeword **c** of the RS code is generated as,

$$c(x) = g(x)u(x), \tag{5}$$

where $u(x) = u_0 + u_1x + u_2x^2 + \cdots + u_{k-1}x^{k-1}$ is the polynomial of the information sequence m, $u_i \in \mathrm{GF}(2^m)$ for $i = 0, 1, \ldots, k-1$. Moreover, the polynomial $c(x)$ of systematic codeword **c** is obtained as,

$$c(x) = x^{n-k}u(x) + p(x), \tag{6}$$

where $p(x) = p_0 + p_1x + p_2x^2 + \cdots + p_{n-k-1}x^{n-k-1}(p_i \in \mathrm{GF}(2^m), i = 0, 1, \ldots, n-k-1)$ denotes the parity-check polynomial which can be computed by the polynomial division as,

$$p(x) = x^{n-k}u(x)/g(x). \tag{7}$$

2.2.3 Decoding of RS codes

Consider a (n, k) RS code with the symbols from GF(q). Suppose that a codeword $c(x) = c_0 + c_1x + \cdots + c_{n-1}x^{n-1}$ is transmitted, and the transmission error result in the following received vector $r(x) = r_0 + r_1x + \cdots + r_{n-1}x^{n-1}$. Let $e(x) = e_0 + e_1x + \cdots + e_{n-1}x^{n-1}$ be the error pattern which have relationship with $c(x)$ and $r(x)$ as,

$$e(x) = r(x) - c(x). \tag{8}$$

Assume that error pattern $e(x)$ contains τ errors (nonzero components) at locations $x^{j_1}, x^{j_2}, \ldots, x^{j_\tau}$, where $0 \le j_1 < j_2 < \cdots < j_\tau \le n-1$. Then,

$$e(x) = e_{j_1}x^{j_1} + e_{j_2}x^{j_2} + \cdots + e_{j_\tau}x^{j_\tau} \tag{9}$$

where x^{j_i} denotes error-location and e_{j_i} is error values, $1 \le i \le \tau$. And the specific decoding steps are given as follows,

Step 1. Compute the syndrome. The syndrome is a $2t$-tuple vector as,

$$\begin{aligned}\mathbf{S} &= (S_1, S_2, \dots, S_{2t}) = \mathbf{r} \cdot H^T \\ &= [r_0, r_1, \dots, r_{n-1}] \cdot \begin{bmatrix} 1 & 1 & 1 & \cdots & 1 \\ \alpha & \alpha^2 & \alpha^3 & \cdots & \alpha^{2t} \\ \alpha^2 & (\alpha^2)^2 & (\alpha^3)^2 & \cdots & (\alpha^{2t})^2 \\ \vdots & \vdots & \vdots & \ddots & \vdots \\ \alpha^{n-1} & (\alpha^2)^{n-1} & (\alpha^3)^{n-1} & \cdots & (\alpha^{2t})^{n-1} \end{bmatrix}.\end{aligned} \tag{10}$$

Evidently, $S_i = r(\alpha^i)\,(1 \le i \le 2t)$.

Step 2. Determined the error-location polynomial $\sigma(x)$ and the error value evaluator $Z_0(x)$ based on Euclidean algorithm.

(1) From Eq. (8) and (10), we obtain,

$$S_i = r(\alpha^i) = e(\alpha^i) + c(\alpha^i) = e(\alpha^i). \tag{11}$$

From Eq. (9), all $2t$ syndromes are obtained,

$$\begin{aligned} S_1 &= e_{j_1}(\alpha^{j_1})^1 + e_{j_2}(\alpha^{j_2})^1 + \cdots + e_{j_\tau}(\alpha^{j_\tau})^1, \\ S_2 &= e_{j_1}(\alpha^{j_1})^2 + e_{j_2}(\alpha^{j_2})^2 + \cdots + e_{j_\tau}(\alpha^{j_\tau})^2, \\ &\vdots \\ S_{2t} &= e_{j_1}(\alpha^{j_1})^{2t} + e_{j_2}(\alpha^{j_2})^{2t} + \cdots + e_{j_\tau}(\alpha^{j_\tau})^{2t}, \end{aligned} \tag{12}$$

where α^{j_i} is called the error location number and e_{j_i} is the error value $(1 \le i \le 2t)$. Let $\beta_i \triangleq \alpha^{j_1}$, $\delta_i \triangleq e_{j_i}$, Eq. (12) can be simplified as,

$$\begin{aligned} S_1 &= \delta_1\beta_1 + \delta_2\beta_2 + \cdots + \delta_\tau\beta_\tau, \\ S_2 &= \delta_1\beta_1^2 + \delta_2\beta_2^2 + \cdots + \delta_\tau\beta_\tau^2, \\ &\vdots \\ S_{2t} &= \delta_1\beta_1^{2t} + \delta_2\beta_2^{2t} + \cdots + \delta_\tau\beta_\tau^{2t}. \end{aligned} \tag{13}$$

(2) To solve these $2t$ equations, the error-location polynomial is firstly defined as:

$$\begin{aligned} \sigma(x) &= (1 - \beta_1 x)(1 - \beta_2 x)\cdots(1 - \beta_\tau x) \\ &= \sigma_0 + \sigma_1 x + \sigma_2 x^2 + \cdots + \sigma_\tau x^\tau. \end{aligned} \tag{14}$$

The roots of $\sigma(x)$ are $\beta_1^{-1}, \beta_2^{-1}, \dots, \beta_\tau^{-1}$, which are the inverses of the error-location numbers [11].

(3) Define error-value evaluator $Z_0(x)$. Firstly, the syndrome polynomial $S(x)$ is defined as,

$$S(x) \triangleq S_1 + S_2 x + S_3 x^2 + \cdots + S_{2t} x^{2t-1} + S_{2t+1} x^{2t} + \cdots = \sum_{j=1}^{\infty} S_j x^{j-1}. \tag{15}$$

Then, $S(x)$ can be further simplified as,

$$S(x) = \sum_{j=1}^{\infty} x^{j-1} \sum_{l=1}^{\tau} \delta_l \beta_l^j = \sum_{l=1}^{\tau} \delta_l \beta_l \sum_{j=1}^{\infty} (x\delta_l)^{j-1}. \tag{16}$$

Since $\frac{1}{1-\beta_l x} = \sum_{j=1}^{\infty}(x\beta_l)^{j-1}$, Thus Eq. (16) comes to,

$$S(x) = \sum_{l=1}^{\tau} \frac{\delta_l \beta_l}{1-\beta_l x}. \tag{17}$$

Then, we have,

$$\begin{aligned}&\sigma(x)S(x) = (1+\sigma_1 x + \cdots + \sigma_\tau x^\tau)(S_1 + S_2 x + S_3 x^2 + \cdots) = S_1 + (S_2 + \sigma_1 S_1)\\&+(S_3 + \sigma_1 S_2 + \sigma_2 S_1)x^2 + \cdots + (S_{2t} + \sigma_1 S_{2t-1} + \sigma_2 S_{2t-2} + \cdots + \sigma_\tau S_{2t-\tau})x^{2t-1}\\&+(S_2 t + \sigma_1 S_{2t-1} + \sigma_2 S_{2t-2} + \cdots + \sigma_\tau S_{2t-\tau})x^2 t + \cdots.\end{aligned} \tag{18}$$

Therefore,

$$\begin{aligned}Z_0(x) &\triangleq \left\{\prod_{i=1}^{\tau}(1-\beta_i x)\right\} \cdot \left\{\sum_{l=1}^{\tau} \frac{\delta_l \beta_l}{1-\beta_l x}\right\}\\&= \sum_{l=1}^{\tau} \frac{\delta_l \beta_l}{1-\beta_l x} \cdot \prod_{i=1}^{\tau}(1-\beta_i x)\\&= \sum_{l=1}^{\tau} \delta_l \beta_l \cdot \prod_{i=1, i\neq l}^{\tau} (1-\beta_i x).\end{aligned} \tag{19}$$

Step 3. Solve the key equation based on the Euclidean algorithm. In the expansion of $\sigma(x)S(x)$, only the coefficient of the first $2t$ terms (from x^0 to x^{2t}) are known. Let $Z_0(x) = [\sigma(x)S(x)]_{2t}$ denote the first $2t$ terms of $\sigma(x)S(x)$. Then, $\sigma(x)S(x) - [\sigma(x)S(x)]_{2t}$ is divisible by x^{2t}. This simply says that if $\sigma(x)S(x)$ is divided by x^{2t}, the remainder is $Z_0(x)$.

Therefore, we obtain,

$$\sigma(x)S(x) = Z_0(x) \mod x^{2t}, \tag{20}$$

which is called the key equation in decoding BCH code. Thus, the key equation can be expressed in the following forms:

$$\sigma(x)S(x) = Q(x)x^{2t} + Z_0(x) \Rightarrow Z_0(x) = -Q(x)x^{2t} + \sigma(x)S(x). \tag{21}$$

Setting,

$$a(x) = x^{2t}, b(x) = S(x). \tag{22}$$

Then the key equation is exactly in the form given as follows,

$$Z_0(x) = -Q(x)a(x) + \sigma(x)b(x). \tag{23}$$

Therefore, $\sigma(x)$ and $Z_0(x)$ can be found by the Euclidean iterative division algorithm. Let

$$Z_0^{(i)}(x) = r_i(x), \sigma^{(i)}(x) = g_i(x), \gamma^{(i)}(x) = -Q^{(i)}(x) = f_i(x). \tag{24}$$

To find $\sigma(x)$ and $Z_0(x)$, we carry out the iteration process as follows:

(1) Firstly, the initial conditions are given as,

$$\begin{aligned} &Z_0^{(-1)}(x) = x^{2t}\left(a(x) = x^{2t}\right), \\ &Z_0^{(0)}(x) = S(X)(b(x) = S(x)), \\ &\gamma^{(-1)}(x) = \sigma^{(0)}(x) = 1, \\ &\gamma^{(0)}(x) = \sigma^{(-1)}(x) = 0.. \end{aligned} \tag{25}$$

(2) Step i: at the i-th step,

$$\begin{aligned} &Z_0^{(i-2)}(x) = q_1(x)Z_0^{(i-1)}(x) + Z_0^{(i)}(x), \\ \Rightarrow \quad &Z_0^{(i)}(x) = \gamma^{(i)}(x)x^{2t} + \sigma^{(i)}(x)S(x), \end{aligned} \tag{26}$$

where

$$\sigma^{(i)}(x) = \sigma^{(i-2)}(x) - q_i(x)\sigma^{(i-1)}(x), \gamma^{(i)}(x) = \gamma^{(i-2)}(x) - q_i(x)\gamma^{(i-1)}(x). \tag{27}$$

(3) Finally, iteration stops when the iteration reaches a step ρ for which

$$\deg\ Z_0^{(\rho)}(x) < \deg\ \sigma^{(\rho)}(x) \le t. \tag{28}$$

Therefore, $Z_0(x) = Z_0^{(\rho)}, \sigma(x) = \sigma^{(\rho)}$ are obtained.

Step 4. Evaluate error location numbers and error values.

(1) Determine error-location numbers α^{j_i} from $\sigma(x)$. The error-location numbers are the inverse of the roots of $\sigma(x)$.

(2) Determine the error values $\delta_l, 1 \le l \le \tau$ from $Z_0(x)$ and $\sigma(x)$. Subsitituting β_l^{-1} in $Z_0(x)$, then,

$$\begin{aligned} Z_0\left(\beta_l^{-1}\right) &= \sum_{l=1}^{\tau} \delta_l\beta_l \prod_{i=1, i\neq l}^{\tau} \left(1 - \beta_i\beta_l^{-1}\right) \\ &= \delta_l\beta_l \prod_{i=1, i\neq l}^{\tau} \left(1 - \beta_i\beta_l^{-1}\right) \end{aligned} \tag{29}$$

(3) Compute the derivative of $\sigma(x)$ as,

$$\sigma'(x) = \frac{\mathrm{d}}{\mathrm{d}x}\prod_{i=1}^{} \tau(1 - \beta_i x) = -\sum_{l=1}^{\tau} \beta_l \prod_{i=1, i\neq l}^{\tau} (1 - \beta_i x). \tag{30}$$

Moreover, substitute β_l^{-1} in Eq. (30) and obtain,

$$\sigma'\left(\beta_l^{-1}\right) = -\beta_l \prod_{i=1, i\neq l}^{\tau} \left(1 - \beta_i\beta_l^{-1}\right). \tag{31}$$

Hence, the error values δ_l at location $beta_l$ is evaluated as,

$$\delta_l = -\frac{Z_0(\beta_l^{-1})}{\sigma'(\beta_l^{-1})}. \tag{32}$$

The Euclidean decoding algorithm is terminated [12].

Example 2. Consider the triple-error-correcting RS code of length $n = 15$ over $GF(2^4)$, α be a primitive element of $GF(2^4)$ such that $\alpha^4 + \alpha + 1 = 0$. The generator polynomial has $\alpha, \alpha^2, \alpha^3, \alpha^4, \alpha^5, \alpha^6$ as roots; that is,

$$\begin{aligned} g(x) &= (x+\alpha)(x+\alpha^2)(x+\alpha^3)(x+\alpha^4)(x+\alpha^5)(x+\alpha^6) \\ &= \alpha^6 + \alpha^9 x + \alpha^9 x^2 + \alpha^4 x^3 + \alpha^{14} x^4 + \alpha^{10} x^5 + x^6. \end{aligned} \tag{33}$$

Suppose that the codeword of all zero is transmitted, and the received polynomial is $r(x) = \alpha^7 x^3 + \alpha^{11} x^{10}$. The decoding procedures are shown as follows,

Step 1. Compute the syndromes $(S_1, S_2, \ldots, S_6)$. The syndrome components are exhibited as,

$$\begin{aligned} S_1 &= r(\alpha) = \alpha^7 \cdot \alpha^3 + \alpha^{11} \cdot \alpha^{10} = \alpha^7, \\ S_2 &= r(\alpha^2) = \alpha^7 \cdot (\alpha^2)^3 + \alpha^{11} \cdot (\alpha^2)^{10} = \alpha^{12}, \\ S_3 &= r(\alpha^3) = \alpha^7 \cdot (\alpha^3)^3 + \alpha^{11} \cdot (\alpha^3)^{10} = \alpha^6, \\ S_4 &= r(\alpha^4) = \alpha^7 \cdot (\alpha^4)^3 + \alpha^{11} \cdot (\alpha^4)^{10} = \alpha^{12}, \\ S_5 &= r(\alpha^5) = \alpha^7 \cdot (\alpha^5)^3 + \alpha^{11} \cdot (\alpha^5)^{10} = \alpha^{14}, \\ S_6 &= r(\alpha^6) = \alpha^7 \cdot (\alpha^6)^3 + \alpha^{11} \cdot (\alpha^6)^{10} = \alpha^{14}. \end{aligned} \tag{34}$$

The syndrome polynomial is $S(x) = \alpha^7 + \alpha^{12} x + \alpha^6 x^2 + \alpha^{12} x^3 + \alpha^{14} x^4 + \alpha^{14} x^5$.

Step 2. Determine the error-location polynomial $\sigma(x)$ and the error-value evaluator $Z_0(x)$ based on the Euclidean algorithm.

(1) Firstly, the initial conditions are acquired as,

$$\begin{aligned} &Z_0^{(-1)}(x) = x^6, \\ &Z_0^{(0)}(x) = S(X) = \alpha^7 + \alpha^{12} x + \alpha^6 x^2 + \alpha^{12} x^3 + \alpha^{14} x^4 + \alpha^{14} x^5, \\ &\gamma^{(-1)}(x) = \sigma^{(0)}(x) = 1, \\ &\gamma^{(0)}(x) = \sigma^{(-1)}(x) = 0.. \end{aligned} \tag{35}$$

(2) When l = 1, then,

$$\begin{aligned} &Z_0^{(-1)}(x) = q_1(x) Z_0^{(0)}(x) + Z_0^{(1)}(x), \\ \Rightarrow\ &x^6 = q_1(x)(\alpha^7 + \alpha^{12} x + \alpha^6 x^2 + \alpha^{12} x^3 + \alpha^{14} x^4 + \alpha^{14} x^5) + Z_0^{(0)}(x), \\ \Rightarrow\ &q_1(x) = \alpha x + \alpha, Z_0^{(1)}(x) = \alpha^6 x^4 + \alpha^5 x^3 + \alpha^5 x^2 + \alpha^3 x + \alpha^8, \end{aligned} \tag{36}$$

where,

$$\sigma^{(1)}(x) = \sigma^{(\ 1)}(x) - q_1(x)\sigma^{(0)}(x) \Rightarrow \sigma^{(1)}(x) = \alpha x + \alpha \tag{37}$$

(3) When l = 2,

$$\begin{aligned} & Z_0^{(0)}(x) = q_2(x)Z_0^{(1)}(x) + Z_0^{(2)}(x), \\ \Rightarrow \ & \alpha^7 + \alpha^{12}x + \alpha^6x^2 + \alpha^{12}x^3 + \alpha^{14}x^4 + \alpha^{14}x^5 \\ = \ & q_2(x)\left(\alpha^6x^4 + \alpha^5x^3 + \alpha^5x^2 + \alpha^3x + \alpha^8\right) + Z_0^{(2)}(x), \\ \Rightarrow \ & q_2(x) = \alpha^8x + \alpha^{11}, Z_0^{(2)}(x) = \alpha^2x + \alpha^3, \end{aligned} \tag{38}$$

where,

$$\sigma^{(2)}(x) = \sigma^{(0)}(x) - q_2(x)\sigma^{(1)}(x) \Rightarrow \sigma^{(2)}(x) = \alpha^9x^2 + \alpha^8x + \alpha^{11} \tag{39}$$

Observe that $\deg Z_0^{(2)}(x) < \deg\sigma^{(2)}(x) \leq 3 = t$. Hence, The iteration is terminated, and we can acquire,

$$Z_0(x) = Z_0^{(2)}(x) = \alpha^2x + \alpha^3, \sigma(x) = \sigma^{(2)}(x) = \alpha^9x^2 + \alpha^8x + \alpha^{11} \tag{40}$$

Step 3. Evaluate error-location numbers and error values. The all roots of $\sigma(x)$ are α^5 and α^{12}. Then, the error location numbers are $(\alpha^5)^{-1} = \alpha^{10}$, $(\alpha^{12})^{-1} = \alpha^3$. The error values at these locations are

$$\begin{aligned} e_3 &= \frac{-Z_0(\alpha^{-3})}{\sigma'(\alpha^{-3})} = \frac{\alpha^3 + \alpha^2\alpha^{-3}}{\alpha^{11}\alpha^3(1 + \alpha^{10}\alpha^{-3})} = \frac{1}{\alpha^8} = \alpha^7, \\ e_{10} &= \frac{-Z_0(\alpha^{-10})}{\sigma'(\alpha^{-10})} = \frac{\alpha^3 + \alpha^2\alpha^{-10}}{\alpha^{11}\alpha^{10}(1 + \alpha^3\alpha^{-10})} = \frac{\alpha^4}{\alpha^8} = \alpha^{11}. \end{aligned} \tag{41}$$

Step 4. Perform error correction.

Therefore, the error polynomial is $e(x) = \alpha^7x^3 + \alpha^{11}x^{10}$. the decoded coded polynomial is $c'(x) = r(x) - e(x) = (\alpha^7x^3 + \alpha^{11}x^{10}) - (\alpha^7x^3 + \alpha^{11}x^{10}) = \mathbf{0}$, which is all-zero codeword.

The end of Example 2.

3. General distributed RS coded-cooperative systems

Coded cooperative diversity is an efficient technique combining channel coding and cooperative diversity to combat the influence of channel fading and improve the performance of the systems [13]. Generally, the coded cooperation is composed of three terminals, i.e., source, relay, and destination. Hence, the channel codes employed in each terminal are named distributed channel codes. Many distributed channel codes are applied in the coded-cooperative systems. For short-to-medium-length transmission information blocks, the RS channel coding may be a promising candidate which illustrates a superior performance [13–16].

Figure 1 demonstrates the general distributed RS coded-cooperative scheme. Evidently, all three terminals transmit and receive signals through one antenna and the entire transmission requires two-time slots. During time slot-1, the binary information sequence $\mathbf{b}_1$ is first converted to the M-ary symbol vector $\mathbf{u}_1$ of length K_1 over the $\mathrm{GF}(2^M)$. Then, $\mathbf{u}_1$ is encoded by the $RS_1(N, K_1, d_1)$ encoder to obtain the systematic

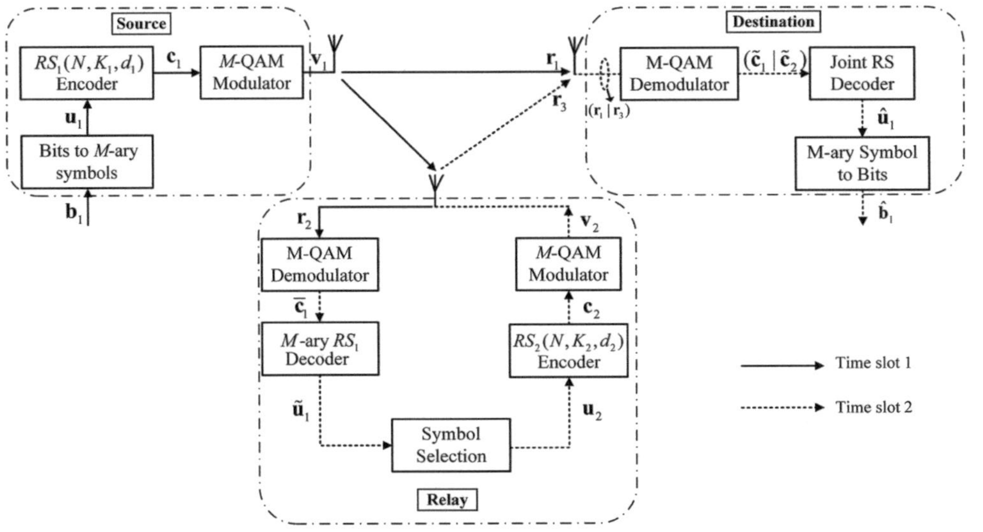

Figure 1.
The system model of the general distributed RS-coded cooperation.

codeword $\mathbf{c}_1$ of length N, where $d_1 = N - K_1 + 1$ and the generator polynomial $g_1(x)$ of RS_1 is given as,

$$g_1(x) = (x - \gamma)(x - \gamma^2)\cdots(x - \gamma^{N-K_1}), \tag{42}$$

where $\gamma^k \in \mathrm{GF}(2^M), k = 0, 1, \ldots, N - K_1$. Then, $\mathbf{c}_1$ is further modulated to the signal $\mathbf{v}_1$ by the M-ary quadrature amplitude modulation (M-QAM). Subsequently, $\mathbf{v}_1 = [v_0, v_1, \ldots, v_{(N-1)}]$ generated at the source is transmitted to the both relay and destination through the respective fading channels where the signals $\mathbf{r}_1 = [r_0^1, r_1^1, \ldots, r_{N-1}^1]$ and $\mathbf{r}_2 = [r_0^2, r_1^2, \ldots, r_{N-1}^2]$ are obtained at the relay and destination, respectively. Moreover, each signal symbol $r_i^j (i = 0, 1, \ldots, N - 1, j = 1, 2)$ is modeled as,

$$r_i^j = h_i^j v_i + n_i^j, \tag{43}$$

where h_i^j is the complex Gaussian variable satisfying zero mean and $1/2$ variance per dimension, and n_i^j represents the complex Gaussian variable with zero mean and $N_0/2$-variance per dimension. Note that N_0 denotes the power spectral density (PSD) of the noise.

During time slot 2, $\mathbf{r}_2$ is demodulated and decoded subsequently to obtain the estimated information sequence $\tilde{\mathbf{u}}_1$. If the source-to-relay channel is ideal, then, $\tilde{\mathbf{u}}_1 = \mathbf{u}_1$. For the system, the information symbols at the relay are only from the source. Therefore, the K_2 symbols are simply chosen from $\tilde{\mathbf{u}}_1$ of length N through the 'Symbol Selection' block. Note that different selection patterns contribute to a different minimum distance of the resultant code at the destination and further affect the overall performance of the RS coded-cooperative scheme, which will be elaborated on in the next section. After that, the selected message vector $\mathbf{u}_2$ is also encoded by the $RS_2(N, K_2, d_2)$ to acquire the c_2, where $d_2 = N - K_2$ and the generator polynomial $g_2(x)$ of RS_2 is provided as,

$$g_1(x) = (x - \gamma)(x - \gamma^2)\cdots(x - \gamma^{N-K_2}), \tag{44}$$

Similarly, the codeword $\mathbf{c}_2$ is modulated by an M-QAM modulator and further transmitted to the destination. The received signal $\mathbf{r}_3$ is also modeled similarly to Eq. (43).

At the destination, the obtained signals $\mathbf{r}_1$ and $\mathbf{r}_3$ are concatenated is series as,

$$\mathbf{r} = (\mathbf{r}_1|\mathbf{r}_3), \tag{45}$$

where '|' denotes that the two signals are conjunct in series during two-time slots. Following that, r passes to the 'M-QAM Demodulator' block to get the joint demodulated message sequence $(\tilde{\mathbf{c}}_1|\tilde{\mathbf{c}}_2)$ and then decoded by the joint RS decoding algorithm that will be introduced in detailed later. Finally, the estimated information sequence $\hat{\mathbf{u}}_1$ is transformed to the extensive bit sequence $\hat{\mathbf{b}}_1$.

4. The optimized codes resulted at destination by proper selection at relay

The different relay selection patterns determine the different minimum distance of the final joint code at the destination, which influences the performance of the system. Therefore, we need to consider the proper selection approach at the relay to capture the resulting code with a minimum distance as large as possible. The following will introduce two proper selection approaches, detailed content can refer to [13].

Obviously, we should consider the worst-case scenario and aim to avoid as many of them as possible. Since the minimum weight of the code at source is already determined as d_1, only the minimum weight of the codeword selected by the relay needs to be considered. Firstly, some nomenclatures are described below before providing design steps:

1. The first scenario is expressed as the minimum weights of code is $wt(\mathbf{c}_1) = d_1$, $wt(\mathbf{c}_2) = 0$ for the source and relay, respectively, resulting in the final code at the destination has the minimum free distance $d_3^{(1)} = d_1$, which is the worst case.

2. The second scenario is described as $wt(c_1) = d_1$ and $wt(c_2) = d_2$. Hence the minimum weight of the final codeword is $d_3^{(2)} = d_1 + d_2$ that is the second-worst case.

3. The third scenario is the weight of the resultant code $d_3^{(3)}$ is greater than $d_3^{(2)}$ at the destination.

4. Define w_1, w_2 and w_3 as the number of times three scenarios occur, respectively.

4.1 Exhaustive search approach

The exhaustive search approach is performed for all information sequences with the weight $0 < wt(\mathbf{u}_1) \leq d_1$ that may be encoded to the codeword with the weight d_1. The preceding are the particular steps of this approach.

1. Define the set $\psi = \{\mathbf{u}_1 | wt(c_1) = d_1\}$ to store the information sequence $\mathbf{u}_1$ that generate the exactly the codeword with weight d_1.

2. Determine the set $\phi = \{\xi_g\}$ which stores all selection patterns $\xi_g = [\xi_1, \xi_2, \dots, \xi_{K_2}]$, where $\xi_i \in 1, 2, \dots, K_2, g = 1, 2, \xi, L$ and L is given as,

$$L = K_1{}_{K_2} = \frac{K_1!}{K_2!(K_1 - K_2)!}. \tag{46}$$

3. For each selection pattern ξ_g, determine the value of w_1. If $|\Gamma| = 1$, Moreover, save the selection patterns corresponding to the $\min(w_1)$ to the set Γ. If then skip step 6 otherwise come to the next step, where $|\cdot|$ denotes the cardinality of the set.

4. From the set Γ, determine the selection patterns ξ_g that correspond to the $\min(w_2)$ and are stored to the set Ω. Similarly, if $|\Omega| = 1$, proceed to step 6, else move to the next step.

5. Determine the selection patterns ξ_g corresponding to $\min(w_3)$ from the set Ω and are further saved in the set Ψ. If $|\Psi| = 1$, then, come to step 6, otherwise add the $wt(\mathbf{c}_2)$ by 1 and move on to step 5 until $|\Psi| = 1$.

6. The optimized selection pattern $\xi^{(ES)} = \xi_g$ is captured. The selection is terminated.

Example 3. In the distributed RS-coded cooperative system, consider the $RS_1(15,11,5)$ and $RS_2(15,7,9)$ are employed in the source and relay, respectively. The symbol elements of the RS_1 and RS_2 are chosen from $GF(2^4)$ shown in **Table 1**. The exhaustive search for selecting the information symbol of $K_2 = 7$ from $K_1 = 11$ is demonstrated below.

1. Find all information sequences $\mathbf{u}_1$ that generate the codewords $\mathbf{c}_1$ with weight $d_1 = 5$. And store them to the set ψ. By numerical simulation, $|\psi| = 45045$.

2. Store all selection patterns $\xi_g = [\xi_1, \xi_2, \dots, \xi_7]$ in the set ϕ. And calculate $|\phi| = L = 330$.

3. Through simulation, $\min(w_1)$ and its corresponding selection patterns are obtained and saved in the set Γ as exhibited in **Table 2**. Since $|\Gamma| = 4 \neq 1$, then come to the next step.

No.	Selection pattern	w_1	w_2
1	[456891011]	840	17,010
2	[457891011]	840	17,280
3	[467891011]	840	17,535
4	[567891011]	840	16,635

Table 2.
The procedure of exhaustive search approach to obtain an optimized selection pattern.

4. For the four selection patterns, determine the $\min(w_2) = 16635$ that corresponding to a selection pattern $\xi_g = [5,6,7,8,9,10,11]$. Thus, $|\Omega| = 1$ and the optimized $\xi^{(ES)} = [5,6,7,8,9,10,11]$ is determined.

The end of Example 3.

4.2 Partial search approach

The exhaustive search approach can choose the optimal selection pattern with the final codeword at the destination having a better weight distribution. However, the complexity of determining the information sequence set ψ and the selection pattern set ϕ increases rapidly when the information length and code length become large. Therefore, we need to consider a low-complexity search approach, i.e., a partial search approach [16]. This approach reduces the search range of the information sequences and the scope of the selection patterns.

First, divide the information positions into two parts illustrated in **Figure 2**. Case (a): the first part is greater than the other part one symbol. Case (b): the last part is greater than the first part symbol. In two cases, make sure the symmetric structure of the K_1 information symbols. Hence, it is reasonable to position the information symbols appropriately. Note that the message sequence generating the codeword with the weight d_1 has at least $\theta = K_1 - \min(K_1, d_1)$ zero symbols. Thus, we focus on selecting the distribution positions of the θ zero symbols and K_2 selection pattern.

1. Determine the distribution positions of the θ zero symbols. For case (a), take $\varepsilon(\lceil \theta/2 \rceil \leq \varepsilon\theta \min(\lceil K_1/2 \rceil, \theta)$ zero symbols set in the first part randomly, and the other $\theta - \varepsilon$ distribute in the last part uniquely. For case (b), ε zero symbols are uniquely assigned in the first part and the remaining $\theta - \varepsilon$ zero symbols are randomly set in the last part, where $\lceil \cdot \rceil$ represent ceil operation. Consider two cases, the set $\overline{\psi}$ that stores partial information sequences generating the codeword with d_1 is determined.

2. Determine the selection positions of K_2 information symbols from the K_1 positions. For case (a), randomly choose $\zeta(\lceil K_2/2 \rceil \leq \zeta \leq \min(\lceil K_1/2 \rceil, K_2))$ positions out of the first part and the left $K_2 - \zeta$ positions are fixed at the last part. For case (b), select ζ positions randomly from the last part and the other $K_2 - \zeta$ positions are uniquely chosen from the first parts. Hence, the reduced selection patterns ξ_g are stored in the set $\overline{\phi}$.

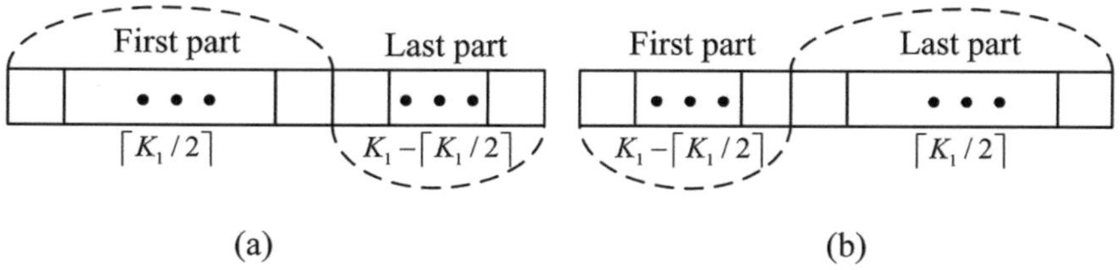

Figure 2.
The symmetric division structure of the positions of K_1 information symbols, case (a) one more symbol in the first part, case (b) one more symbol in the last part.

Based on the reduced sets $\overline{\psi}$ and $\overline{\phi}$, the subsequent steps are same as the Step 3–6 of the exhaustive search approach.

Example 4. This example uses the same codes as Example 4.1. Evidently, the information sequence that can be encoded to the codeword with weight 5 includes at least 6 zero-symbols. The division structure of the partial search approach is shown in **Figure 3**.

1. Determine the distribution positions of the 6 zero symbols. For case (a), take $\varepsilon(\varepsilon = 3,4,5,6)$ zero symbols set in the first 6 positions randomly, and the other $6 - \varepsilon$ distribute in the last 5 positions uniquely. For case (b), ε zero symbols are uniquely assigned in the first part and the remaining $6 - \varepsilon$ zero symbols are randomly set in the last part. Consider two cases, the set $\overline{\psi}$ is determined and $|\psi| = 24075$.

2. Determine the selection positions of 7 information symbols from the 11 positions. For case (a), randomly choose $\zeta(\zeta = 4,5,6)$ positions out of the first 6 positions, and the left $7 - \zeta$ positions are fixed at the last 5 positions. For case (b), select ζ positions randomly from the last 6 positions, and the other $7 - \zeta$ positions are uniquely chosen from the first 5 positions. Hence, the set $\overline{\phi}$ of the partial selection pattern is determined, and $|\overline{\phi}| = 44$.

3. Through simulation, $\min w_1 = 360$ and its corresponding selection patterns are stored in the set Γ as demonstrated in **Table 3**. Since $|\Gamma| = 3 \neq 1$, then come to the next step.

4. For the three selection patterns, determine $\min(w_2)$ that corresponds to two selection patterns. Thus, $|\Omega| = 2 \neq 1$, go to the next step.

5. Obtain the $\min(w_3) = 6540$ and corresponding selection pattern from the set Ω and are further saved in the set Ψ. Since $|\Psi| = 1$, then, the optimized selection pattern $\xi^{(PS)} = [1\,2\,3\,6\,9\,10\,11]$ is acquired. The partial search stops.

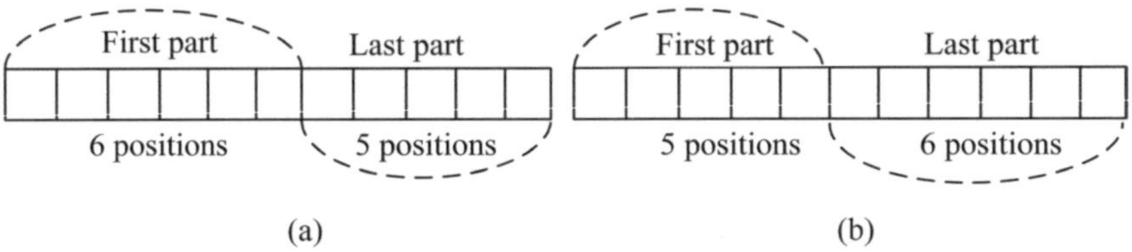

Figure 3.
The symmetric division structure of the positions of 11 information symbols, case (a) 6 symbols in the first part, case (b) 6 symbols in the last part.

No.	Selection pattern	w_1	w_2	w_3
1	[1 2 3 4 9 10 11]	360	10,035	6615
2	[1 2 3 6 9 10 11]	360	10,035	6540
3	[1 2 3 8 9 10 11]	360	10,035	– – – –

Table 3.
The procedure of partial search approach to obtain an optimized selection pattern.

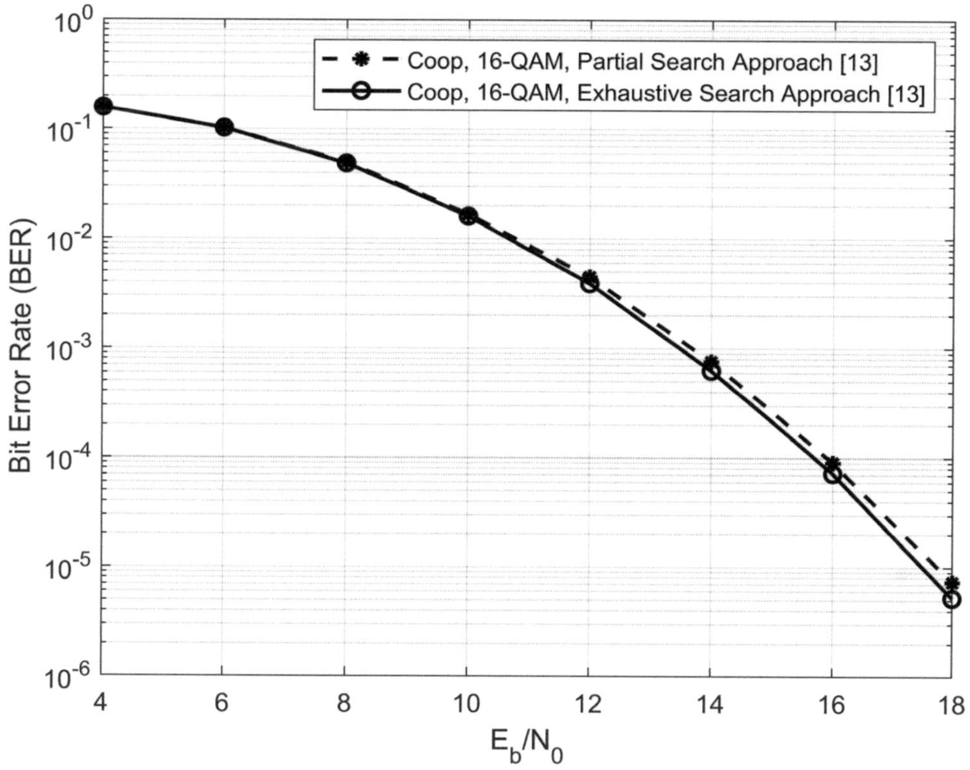

Figure 4.
The BER performance comparison of the distributed RS-coded cooperative scheme with two selection approaches at the relay over the fast-fading channel.

The end of Example 4.

Based on Examples 3 and 4, the BER performance of the distributed RS-coded cooperative scheme over the Rayleigh fast-fading channel employing the exhaustive search and partial search is exhibited in **Figure 4** where the 16-QAM modulation is employed and the source-to-relay channel is ideal. The result reveals that the scheme with two different approaches illustrates almost identical performance, which further shows the feasibility of the reduced-complexity approach. More simulation results can refer to [13].

4.3 Complexity comparisons

First, the complexity comparisons of the two search approaches are listed in **Table 4**, where $(\lambda_1^+, \lambda_1^\times)$ and $(\lambda_2^+, \lambda_2^\times)$ represent the number of the operations of the addition and the multiplication required to encode the information sequence from the set ψ and $\overline{\psi}$ at the source and relay, respectively, and λ_{total} denotes the total operations.

5. Joint decoding algorithms and error performance analysis

The section introduces the two joint decoding algorithms, namely, the naive algorithm and the smart algorithm. The two decoding algorithms may enhance the overall

Approaches	Operations	$(\lambda_1^+, \lambda_1^\times)$	$(\lambda_2^+, \lambda_2^\times)$	λ_{total}
Exhaustive Search		$(K_1(N-K_1)\lvert\psi\rvert,$	$(K_2\lvert\psi\rvert\lvert\phi\rvert(N-K_2),$	$2\lvert\psi\rvert[NK_1 + NK_2\lvert\phi\rvert$
		$K_1(N-K_1)\lvert\psi\rvert)$	$K_2\lvert\psi\rvert\lvert\phi\rvert(N-K_2))$	$-(K_1)^2 - (K_2)^2\lvert\phi\rvert]$
Partial Search		$(K_1(N-K_1)\lvert\overline{\psi}\rvert,$	$K_1(N-K_1)\lvert\overline{\psi}\rvert)$	$(K_2\lvert\overline{\psi}\rvert\lvert\overline{\phi}\rvert(N-K_2),$
		$K_2\lvert\overline{\psi}\rvert\lvert\overline{\phi}\rvert(N-K_2))$	$2\lvert\overline{\psi}\rvert[NK_1 + NK_2\lvert\overline{\phi}\rvert$	$-(K_1)^2 - (K_2)^2\lvert\overline{\phi}\rvert]$

Table 4.
Complexity comparisons of two approaches.

performance by making full advantage of the two signals from the source and relay, respectively.

5.1 Nave decoding algorithmm

The detailed steps for the naive algorithm are listed as follows:

1. For the received demodulated signal $(\tilde{\mathbf{c}}_1|\tilde{\mathbf{c}}_2)$, $\tilde{\mathbf{c}}_1$ and $\tilde{\mathbf{c}}_2$ are decoded by RS_1 and RS_2 decoders, respectively, to acquire the estimated information sequences $\mathbf{u}_1'$ and $\mathbf{u}_2'$.

2. Determine the SNR cross-point of the RS_1 and RS_2 point-to-point coding scheme over the fast-fading channel, denoted η.

3. If SNR $\leq \eta$, $\hat{\mathbf{u}}_1 = \mathbf{u}_1'$ due to the better performance of RS_1 code than that of RS_2 code at the low SNRs. Otherwise, $\mathbf{u}_2'$ replaces $\mathbf{u}_1'$ at the corresponding selected positions to obtain a re-combined $\ddot{\mathbf{u}}_1$, then, $\hat{\mathbf{u}}_1 = \ddot{\mathbf{u}}_1$. This is because the RS_2 code with more parity-check symbols outperforms the RS_1 code at high SNRs. Finally, the estimated sequence $\hat{\mathbf{u}}_1$ is obtained.

5.2 Smart decoding algorithm

The specific steps for the smart algorithm are described below:

1. For the received demodulated signal $(\tilde{\mathbf{c}}_1|\tilde{\mathbf{c}}_2)$, only decode the last part $\tilde{\mathbf{c}}_2$ to get the systematic non-binary message sequence $\mathbf{u}_2'$.

2. For the first part $\tilde{\mathbf{c}}_1$ comprising of the check-parity sequence $\tilde{\mathbf{p}}_1$ and information sequence $\tilde{\mathbf{u}}_1$, replace the non-binary symbols of $\tilde{\mathbf{u}}_1$ with $\mathbf{u}_2'$ in the corresponding K_2 positions to obtain the re-combined sequence $\overline{\mathbf{c}}_1$ due to the reliability of $\mathbf{u}_2'$ than original message symbols.

3. Decode $\overline{\mathbf{c}}_1$ by the RS_1 decoder to acquire the final estimated information sequence $\overline{\mathbf{u}}_1$.

Figure 5 illustrates the BER performance of the distributed RS-coded cooperative scheme under two different decoding algorithms over a fast fading channel, where 16-QAM is applied in the scheme and the partial search approach is employed in the

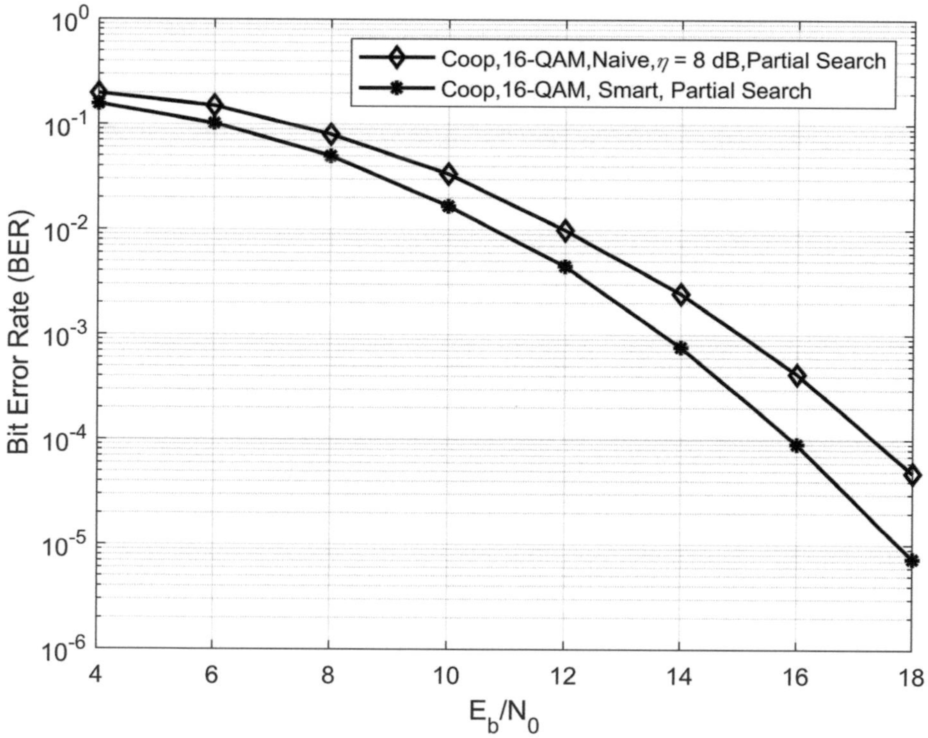

Figure 5.
The performance comparison of the distributed RS-coded cooperative scheme under two different joint decoding algorithms over the fast-fading channel.

relay. From the simulated result, the scheme under the smart decoding algorithm is superior to that of the naive by a gain of over 1.5 dB at $BER \approx 4 \times 10^{-5}$.

5.3 Error performance of distributed RS coded-cooperative systems

This section presents the average error probability (AEP) bound for the distributed RS coded-cooperative scheme over the Rayleigh fast-fading channel. First, the unconditional error probability is provided as follows [5, 17, 18],

$$P_b(E) = \frac{1}{\pi}\int_0^{\pi/2}\left(1+\frac{\Lambda_1}{\sin^2\varphi}\right)^{d_1}\left(1+\frac{\Lambda_2}{\sin^2\varphi}\right)^{d_2} d\varphi, \tag{47}$$

where Λ_1 and Λ_2 denote the average signal-to-noise ratio (SNR) per information bit from the source-to-destination and relay-to-destination links. The integral in Eq. (47) is calculated by the available computer package. Then, the upper bound may be acquired by assuming $\sin^2\varphi = 1$, shown as,

$$P_b(E) \le \frac{1}{2}\left(\frac{1}{1+\Lambda_1}\right)^{d_1}\left(\frac{1}{1+\Lambda_2}\right)^{d_2}, \tag{48}$$

Therefore, based on Eq. (48), the upper bound of the bit error probability P_b is further given as [6],

$$P_b \leq \sum_{\varpi=d_1+d_2}^{N} \frac{J_\varpi}{K_1} P_b(E), \tag{49}$$

where J_ϖ represents a weight enumerating factor for each codeword with weight w which is obtained by exhaustive computer search.

6. Conclusions

The chapter first introduces the encoding and decoding procedure of the BCH codes and RS codes. Then, the system model of the distributed RS-coded cooperation is presented which improves the anti-interference transmission performance of the short-to-medium-length information block. In the scheme, the exhaustive and partial search approaches are introduced and employed in the relay to choose an optimized selection pattern that results in a final code with a better weight distribution at the destination. In addition, two joint decoding algorithms are provided to further enhance the performance and the performance analysis validates the system.

Acknowledgements

The financial assistance provided by the National Natural Science Foundation of China under contract No. 61771241 is acknowledged.

Conflict of interest

The authors declare no conflict of interest.

Author details

Chen Chen and Fengfan Yang*
College of Electronics and Information Engineering, Nanjing University of Aeronautics and Astronautics, Nanjing, China

*Address all correspondence to: yffee@nuaa.edu.cn

References

[1] Alamouti SM. A simple transmit diversity technique for wireless communications. IEEE Journal on Selected Areas in Communications. 1998;**16**(8):1451-1458. DOI: 10.1109/49.730453

[2] Ejaz S, Yang FF. Turbo codes with modified code matched interleaver for coded-cooperation in half-duplex wireless relay networks. Frequenz. 2015; **69**(3–4):171-184. DOI: 10.1515/freq-2014-0072

[3] Wang H, Chen Q. LDPC based network coded cooperation design for multi-way relay networks. IEEE Access. 2019;7:62300-62311. DOI: 10.1109/ACCESS.2019.2915293

[4] Umar YFF, Mughal S. Distributed polar coded single carrier-FDMA based on multilevel construction over multipath channels. Wireless Personal Communications. 2019;**105**(3):835-856. DOI: 10.1007/s11277-019-06124-4

[5] Park J, Kim J. Generator polynomial model-based eye diagram estimation method for Bose-Chaudhuri-Hocquenghem (BCH) code and reed-Solomon (RS) code. IEEE Transactions on Electromagnetic Compatibility. 2020; **62**(1):240-248. DOI: 10.1109/TEMC.2018.2881146

[6] Ejaz S, Yang FF. Jointly optimized reed-uller codes for multilevel multirelay coded-cooperative VANETS. IEEE Transactions on Vehicular Technology. 2017;**66**(5):4017-4028. DOI: 10.1109/TVT.2016.2604320

[7] Gong B, Ding C, Li C. The dual codes of several classes of BCH codes. IEEE Transactions on Information Theory. 2022;**68**(2):953-964. DOI: 10.1109/TIT.2021.3125933

[8] Guruswami V, Sudan M. Improved decoding of Reed-Solomon and algebraic-geometry codes. IEEE Transactions on Information Theory. 1999;**45**(6):1757-1767. DOI: 10.1109/18.782097

[9] Barry JR, Lee EA, Messerschmitt DG. Digital Communication. 3rd ed. Springer US; 2004

[10] Blasco FL, Garrammone G, Liva G. Parallel concatenation of non-binary linear random fountain codes with maximum distance separable codes. IEEE Transactions on Communications. 2013;**61**(10):4067-4075. DOI: 10.1109/TCOMM.2013.090513.120834

[11] Zeh A, Li W. Decoding Reed-Solomon codes up to the Sudan radius with the Euclidean algorithm. In: 2010 International Symposium On Information Theory & Its Applications. 2010. pp. 986-990

[12] Andreas FM. Channel coding and information theory. Wireless Communications. IEEE. 2011;**63**:277-317. DOI: 10.1002/9781119992806.ch14

[13] Guo PC, Yang FF, Zhao CL, Ullah W. Jointly optimized design of distributed Reed-Solomon codes by proper selection in relay. Telecommunication System. 2021;**78**(3):391-403. DOI: 10.1007/s11235-021-00822-w

[14] Halbawi W, Ho T, Yao HY, Duursma I. Distributed Reed-Solomon codes for simple multiple access networks. In: IEEE International Symposium on Information Theory. 2014. pp. 651-655

[15] Zhao C, Yang FF, Waweru DK. Reed-Solomon coded cooperative spatial

modulation based on nested construction for wireless communication. Radioengineering. 2021;**30**(1):172-183. DOI: 10.13164/re.2021.0172

[16] Chen C, Yang FF, Zhao CL, Xu HJ. Distributed reed-Solomon coded cooperative space-time labeling diversity network. Radioengineering. 2022;**4**(96): 496-509. DOI: 10.13164/re.2022.0496

[17] Hunter TE, Nosratinia A. Diversity through coded cooperation. IEEE Transactions on Wireless Communications. 2006;**5**(2):283-289. DOI: 10.1109/TWC.2006.1611050

[18] Simon MK, Alouini M. A unified approach to the performance analysis of digital communication over generalized fading channels. Proceedings of the IEEE. 1998;**86**(9):1860-1877. DOI: 10.1109/5.705532

Chapter 3

Linear Codes from Projective Varieties: A Survey

Rita Vincenti

Abstract

Linear codes can be constructed from classical algebraic varieties or from appropriate subsets of finite geometry by considering projective systems arising from their rational points. This geometric point of view allows to look for linear codes by choosing suitable sets to get immediately length, minimum distance, and spectrum (cf. Lemma 1, Propositions 5, 9, 12, 13, 17). In some cases, it is also possible to build a PD-set or an antiblocking decoding (cf. Propositions 3, 4, 14, Examples of Section 5).

Keywords: finite projective geometry, projective systems, linear codes, quadrics, schubert variety

1. Introduction

In the construction of a code, it would be desirable to have a small length, great dimension, and minimum distance to keep the transmission rate low and to get both much information and the correction of many errors. But it is impossible to improve all the basic parameters at the same time.

The purpose of this article is to collect results on linear codes arising from classical varieties via projective systems, by adopting a geometric point of view. If the basic parameters of such varieties (or of appropriate subsets of points) in a projective space can be calculated directly, then it is possible to know immediately length and minimum distance and sometimes also the weight distribution of the related linear codes.

When the group of automorphisms of a variety is read in the automorphism group of a related code, in some cases, it is possible to construct a PD-set (Permutation Decoding set) and/or an AI-system (Antiblocking system).

A PD-set for a t-error-correcting code C is a set S of automorphisms of the code such that every possible error vector of weight $w \leq t$ can be moved out of the information positions by some member of S (cf. [1, 2]). To apply a PD-set to decode a message refer to Huffman ([3], pp. 1345–1440), where an algorithm is given. The permutation decoding algorithm is more efficient the smaller size of the PD-set. The Gordon (lower) bound on this size is crucial (cf. [2] and [3], p. 1414).

A large automorphism group of a code allows to find a PD-set. Indeed, linear codes defined by projective systems usually have large automorphism group. In [4] is generalized the notion of a PD-set of a code to that of a t-PD-set of an arbitrary permutation set.

An AI-system (Antiblocking system) is a new decoding algorithm developed by Kroll and Vincenti in [5, 6], which is comparable to the permutation decoding algorithm, but more efficient being simpler and faster than the permutation decoding algorithm. The existence of a PD-set implies usually also the existence of an AI-system of the same size. But there also may exist AI-systems that are not derived from PD-sets and which are smaller than the known PD-sets. By comparing the two decoding algorithms, it is clear that the antiblocking decoding needs less computing steps than the permutation decoding, even if the size of both systems is the same. Moreover, the antiblocking decoding may be applied even if there does not exist a PD-set or if there exists only a PD-set of very large size and an AI-system of smaller size. For the technique to find small AI-systems, refer to [5], where some properties of antiblocking systems are established.

Codes related to quadrics in the 3-dimensional finite projective spaces $PG(3,q)$ mark the way for the next examples. The geometries of the plane sections of the quadrics over a conic are well known, as well as their automorphism groups. In Section 3, PD-sets for $q = 3$ for all three cases are presented. For the elliptic and hyperbolic quadric, also examples for $q = 4$ are given. These results say that the corresponding codes admit PD-sets S. The size $|S|$ is minimal in some cases. For the hyperbolic quadric also a 5-AI-system is shown, while the Gordon bound is 6 (cf. Propositions 3, 4).

In Section 4, we will refer to the construction of linear codes arising, respectively, from the Grassmannian of the lines of the third dimension (that is, the Klein quadric) and from the Schubert variety of $PG(5,q)$ (cf. Proposition 5, 6 and Examples 1, 2).

In Section 5, we consider codes related to what we call the *celtic variety*, that is, the ruled rational normal surface V_2^3 of order three in $PG(4,q)$ (cf. Propositions 13). Examples of PD-sets for $q = 3$ and $q = 4$ are given in Proposition 14.

In the last Section, results concerning projective systems and codes related to *ruled sets* are collected (cf. Proposition 17 and Examples 3, 4, 5). In the examples, the weight distribution of each code is also shown.

The title of each section refers only to the varieties of which the related codes are described there.

2. Codes and projective systems

Let $F = GF(q)$ be a finite field, $q = p^s$, p prime, denote by F^n the n-dimensional vector space over F.

A *linear* $[n,k]_q$*-code* C *of length* n is a k-dimensional subspace of the vector space F^n.

For $t \geq 1$ the t-th *higher weight* of C (cf. Wei [7]) is defined by

$$d_t = d_t(C) = \min\{\|D\| \text{ for all } D < C, \ \dim D = t\}, \tag{1}$$

where D is a subspace of C and $\|D\|$ is the number of indices i such that there exists $v \in D$ with $v_i \neq 0$.

The first parameter $d_1 = d_1(C)$ is the *Hamming distance* d, that is, the classical *minimum distance* (or, *minimum weight*) of C.

The code C has *genus* at most $g \geq 0$ if $k + d_1 \geq n + 1 - g$.

Sometimes an $[n,k]_q$-code C of minimum distance d is denoted $[n,k,d]_q$-code.

Let $P^{k-1} = PrF^k = PG(q-1, q)$ denote the $(k-1)$-dimensional Galois projective space over the field F, $k \geq 3$ with point set $\mathcal{P}$ and line set $\mathfrak{L}$. Denote $\mathfrak{T}$ the set of all subspaces of $PG(k-1, q)$, $\mathfrak{H}$ the set of the hyperplanes of $PG(k-1, q)$.

The *incidence hull* of a subset $X \subset \mathcal{P}$ is denoted by $\overline{X}$. Thus the joining line of two points $X, Y \in \mathcal{P}$ is $\overline{X, Y} := \overline{\{X, Y\}}$.

An *$[n, k]$-projective system* $\mathcal{X}$ of P^{k-1} is a collection of n points. $\mathcal{X}$ is *non-degenerate* if its n points are not contained in any hyperplane.

From now on assume that $\mathcal{X}$ consists of n distinct points of rank k.

A *standard matrix* $\mathcal{M}$ can be constructed as follows: for each of the n points of $\mathcal{X}$ choose a generating vector, such n vectors are the rows of $\mathcal{M}$. Let $C_{\mathcal{X}}$ be the linear code having $\mathcal{M}^t$ as a generatrix matrix. The code $C_{\mathcal{X}}$ *is the k-dimensional subspace of* F^n which is the *image of the mapping* $(F^k)^* \twoheadrightarrow F^n$ *from the dual k-dimensional space* $(F^k)^*$ *onto* F^n *that calculates every linear form over the points of* $\mathcal{X}$. Therefore the length n of a codeword $C_{\mathcal{X}}$ is the cardinality of $\mathcal{X}$, and the dimension of $C_{\mathcal{X}}$ is k.

An *automorphism* of the code C is a weight-preserving linear automorphism (cf. [4], Section 2).

The equivalences among $[n, k, d]_q$-codes are the restrictions of the automorphisms of F^n represented by *monomial matrices*, where a monomial matrix is the product of a permutation matrix and a diagonal matrix (for the basic concepts of coding theory see for example [3]). To any subset representing the n points of $\mathcal{X}$ is associated a *linear $[n, k, d]$-code*, any two such $[n, k, d]_q$-linear codes are *equivalent*.

A natural 1–1 correspondence connects the equivalence classes of a non-degenerate *$[n, k]_q$-projective system* $\mathcal{X}$ with a non-degenerate *$[n, k]_q$-code* $C_{\mathcal{X}}$. If $\mathcal{X}$ is an $[n, k]_q$-projective system and $C_{\mathcal{X}}$ is a corresponding code, then the non-zero codewords of $C_{\mathcal{X}}$ correspond to hyperplanes $\mathfrak{H}$ of P^{k-1}, up to a non-zero factor, the correspondence preserving the parameters n, k, d_t. Therefore, the weight of a codeword $\mathbf{c}$ corresponding to the hyperplane $H_{\mathbf{c}}$ is the number of points of $\mathcal{X} \setminus H_{\mathbf{c}}$ so that the minimum weight d of the code $C_{\mathcal{X}}$ is $d = |\mathcal{X}| - max\{|\mathcal{X} \cap H| \; |H \in \mathfrak{H}\}$.

A linear code C having d as minimum weight is an s-error-correcting code for all $s \leq \lfloor \frac{d-1}{2} \rfloor$, and $t = \lfloor \frac{d-1}{2} \rfloor$ is the *error-correcting capability of C*.

Subcodes D of C of dimension r correspond to subspaces of codimension r of P^{k-1}. Therefore the higher weights of C are
$d_t = d_t(C) = n - \max\{|\mathcal{X} \cap S| : S < P^{k-1}| \text{ codim } S = t, \}$. In particular,
$d_1 = d_1(C) = n - \max\{|\mathcal{X} \cap H| : H < P^{k-1}| \text{ codim } H = 1, \}$.

The *spectrum* of a projective system $\mathcal{X}$ of P^{k-1} is the set of the following numbers $A_i^{(s)} = |\{S < P^{k-1} : \text{ codim } S = s, |S \cap \mathcal{X}| = n - i\}|$ for all $i = 1, 2, \ldots, n, s = 1, 2, \ldots, k-2$.

Let $H \in \mathfrak{H}$ be a hyperplane. An *intersection number of* $\mathcal{X}$ *(with respect to hyperplanes)* is $|\mathcal{X} \cap H|$. The *type of* $\mathcal{X}$ *with respect to the hyperplanes* is the set $M_{\mathcal{P}}$ of all intersection numbers of $\mathcal{X}$. For $i \in M_{\mathcal{P}}, t_i := \{H \in \mathfrak{H} | \; |\mathcal{X} \cap H| = i\}$ is the total number of hyperplanes providing the intersection number i.

Let $\mathcal{X}$ be a projective system of type $M_{\mathcal{P}}$. Then for $i \in M_{\mathcal{P}}$ *there are* t_i *code words in the related code* $C_{\mathcal{X}}$ *of weight* $|\mathcal{X}| - i$. Analogous definitions can be stated for all subspaces of $\mathfrak{T}$. Therefore, the spectrum of $\mathcal{X}$ induces the *weight distribution of the codewords of* $C_{\mathcal{X}}$.

For the above definitions see also [8–10].

For the definitions of the permutation and the antiblocking decoding and the respective algorithms, see [5, 6], p. 1463.

The following result holds (cf. [8]):

Result A (non-degenerate) projective system of P^k satisfies the following

Gordon bound: (cf. [2, 11]) Let S be a PD-set for a t-error-correcting $[n, k]$-code with redundancy $r = n - k$. Then

$$|S| \geq \left\lceil \frac{n}{r} \left\lceil \frac{n-1}{r-1} \cdots \left\lceil \frac{n-t+1}{r-t+1} \right\rceil \cdots \right\rceil \right\rceil. \tag{2}$$

Following the geometric interpretation of a linear code shown by Tsfasman and Vladut (cf. [12, 13]) many authors studied codes arising from sets of the rational points of an affine or of a projective space.

Denote $\overline{F}$ the algebraic closure of the field $F = GF(q)$.

The geometry $PG(r, q) = P^r$ is considered a sub-geometry of $\overline{PG}(r, q) = \overline{P}^r$, projective geometry over $\overline{F}$. We refer to the points of P^r as the *rational points* of $\overline{P}^r$.

A *variety* V_u^v *of dimension* u *and of order* v of P^r is the set of the rational points of a projective variety $\overline{V}_u^v$ of $\overline{P}^r$ defined by a finite set of polynomials of $F[x_0, \dots, x_r]$.

Choose a coordinate system in P^r so that it is a coordinate system for $\overline{P}^r$ too, denote a point $P \approx (x_0, x_1, \dots, x_r) := \overline{F}^* (x_0, x_1, \dots, x_r), \overline{F}^* = \overline{F} \backslash \{0\}$.

P is a *rational point* if there exists $(x_0, x_1, \dots, x_r) \in F^{r+1}$ such that $P \approx (x_0, x_1, \dots, x_r)$.

For the definition of *projective variety,* the reader can refer to [14, 16].

If $\mathcal{X}$ is a subset of $n = |\mathcal{X}|$ points of P^r, then a subspace of P^r of projective dimension u is denoted by S_u. A variety of dimension u and of order v is denoted V_u^v.

A *t-secant subspace* is a subspace intersecting $\mathcal{X}$ in t points. A *t-secant* is a t-secant line.

A line t is a *tangent* of $\mathcal{X}$ if either t has just one point in common with $\mathcal{X}$ or, each point of t is contained in $\mathcal{X}$. If a tangent line t has just one point in common with $\mathcal{X}$, then t is *a tangent of* $\mathcal{X}$ *at* the point P. A subspace $\mathcal{U}$ is a *component-subspace* if each point of $\mathcal{U}$ lies in $\mathcal{X}$. A line l with the property that each of its points lies in $\mathcal{X}$ is a *component-line*, or simply a *component*. If $m - 1$ is the largest dimension of a component space of $\mathcal{X}$, then m is the *vector index* of $\mathcal{X}$.

3. The quadrics in $PG(3, q)$

In the 3-dimensional geometry $P^3 = PG(3, q)$ over the Galois field $F = GF(q)$ we will consider the *elliptic quadric* $\mathcal{E}_3$, the *hyperbolic quadric* $\mathcal{H}_3$, and the *quadric cone* $\mathcal{Q}_O$ with vertex O and directrix a conic of a plane π, $O \notin \pi$..

Denote $\mathcal{Q}$ the projective system defined by the rational points of a quadric Q. Let C_Q be a related code.

The minimum weights d_t of the projective system $\mathcal{Q}$ consisting of n points are, in such a dimension, $d_t = n - \max\{|\mathcal{Q} \cap S_t|\}, \ t = 1, 2,$

and the spectrum of $\mathcal{Q}$ is

$$A_{n-i}^{(s)} = |\{S_{3-s} < PG(3, q) : \ |S_{3-s} \cap \mathcal{Q}| = i\}|$$

$$i = 1, 2, \dots, n, \ \ s = 1, 2. \tag{3}$$

To construct a linear code related to a quadric, one needs to know all the *intersection numbers* related to it.

From [14] we get

a. the elliptic quadric $\mathcal{E}_3: \ f(x_0, x_1) + x_2x_3 = 0,$ consists of q^2+1 points no three of which are collinear, q^2+1 tangent planes, $q(q^2+1)$ and secant planes;

b. the hyperbolic quadric $\mathcal{H}_3: \ x_0x_1 + x_2x_3 = 0$ consists of $(q+1)^2 = q^2+2q+1$ points on $2(q+1)$ (component) lines, each point on two lines, $q(q^2-1)$ secant planes through conics;

c. the cone $\mathcal{Q}_O: \ x_0^2 + x_1x_4 = 0,$ consists of $q(q+1)+1 = q^2+q+1$ points on the $q+1$ (component) lines projecting from the point $\mathcal{O}$ a conic directrix.

Let H be a plane of $\mathfrak{H}$. If $\mathcal{Q}$ is elliptic, then $\mathcal{Q} \cap H$ is a point or a conic. If $\mathcal{Q}$ is hyperbolic, then $\mathcal{Q} \cap H$ is the union of two lines or a conic. If $\mathcal{Q}$ is a cone, then $\mathcal{Q} \cap H$ is the vertex $\mathcal{O}$ or a line or the union of two lines or a conic.

From [14] and from above the basic parameters and the spectrum of the projective system $\mathcal{Q}$ can be shown.

Lemma 1 (1) *If* $\mathcal{Q} = \mathcal{E}_3$ *then* $n = q^2+1, k = 4, d_1 = q^2 - q, d_2 = q^2 - 1;$

$A^{(1)}_{n-(q+1)} = q(q^2+1), A^{(1)}_{n-1} = q^2+1, A^{(2)}_{n-2} = q^2\frac{(q^2+1)}{2}, A^{(2)}_{n-1} = (q+1)(q^2+1),$
$A^{(2)}_{n} = q^2\frac{(q^2+1)}{2}$ *and no other parameter is different from zero.*

(2) *If* $\mathcal{Q} = \mathcal{H}_3$ *then* $n = (q+1)^2, k = 4, d_1 = q^2, d_2 = q^2+q;$

$$A^{(1)}_{n-(2q+1)} = (q+1)^2, A^{(1)}_{n-(q+1)} = q(q^2-1); \tag{4}$$

$A^{(2)}_{n-(q+1)} = 2(q+1), A^{(2)}_{n-2} = q^2\frac{(q+1)^2}{2}, A^{(2)}_{n-1} = (q+1)(q^2-1), A^{(2)}_{n} = \frac{q(q-1)^2}{2}$ *and no other parameter is different from zero.*

(3) *If* $\mathcal{Q} = \mathcal{Q}_O$ *then* $n = q^2+q+1, k = 4, d_1 = q^2 - q, d_2 = q^2;$

$$A^{(1)}_{n-(2q+1)} = \frac{q(q+1)}{2}, A^{(1)}_{n-(q+1)} = q^3+q+1, A^{(1)}_{n-1} = \frac{q(q-1)}{2}; \tag{5}$$

$$A^{(2)}_{n-(q+1)} = q+1, A^{(2)}_{n-2} = \frac{q^3(q+1)}{2}, \tag{6}$$

$A^{(2)}_{n-1} = q^3+q^2, A^{(2)}_{n} = \frac{q^3(q-1)}{2}$ *and no other parameter is different from zero.*

Then is proved the following (cf. Proposition 5 of [4]).

Proposition 2 (1) *If* $\mathcal{Q} = \mathcal{E}_3$*, then* $\mathcal{Q}$ *is a* $[q^2+1, 4, q(q-1)]_q$*-projective system. 2) If* $\mathcal{Q} = \mathcal{H}_3$*, then* $\mathcal{Q}$ *is a* $\left[(q+1)^2, 4, q^2\right]_q$*-projective system.* 3) If $\mathcal{Q} = \mathcal{Q}_O$*, then* $\mathcal{Q}$ *is a* $[q(q+1), 4, q(q-1)]_q$*-projective system.*

The following propositions supply examples for $q = 3$ and $q = 4$. Denote $\mathcal{Q}$ a quadric in $PG(3,3)$ and C_Q a related code. From [4], Proposition 11 we get

Proposition 3 (1) *If* $\mathcal{Q} = \mathcal{E}_3$*, then* C_Q *is a 2-error-correcting* $[10,4,6]_3$*-code admitting a PD-set S of minimum size* 4.

(2) *If* $\mathcal{Q} = \mathcal{H}_3$, *then* C_Q *is a 4-error-correcting* $[16,4,9]_3$*-code admitting a PD-set of size* 8.

(3) *If* $\mathcal{Q}' = \mathcal{Q}_O \backslash \{O\}$, *then a related code* C'_Q *is a 2-error-correcting* $[12,4,6]_3$*-code admitting a PD-set S of minimum size* 3.

Denote $\mathcal{Q}$ a quadric in $PG(3,4)$ and C_Q a related code. From [4], Propositions 12, 13 and from the Example of [6], p.1464, we get

Proposition 4 (*a*) *If* $\mathcal{Q}$ *is an elliptic quadric, then* C_Q *is a 5-error-correcting-* $[17,4,12]_4$*-code admitting a PD-set* $\mathcal{S}$ *of size* 16.

Let $\mathcal{Q}$ be a hyperbolic quadric. Then

(b_1) C_Q *is a* $[25,4,16]_4$*-code admitting a PD-set* $\mathcal{S}$ *of size* 20 *and a 5-AI-system* $\mathcal{A}$.

(b_2) *If* $P \in \mathcal{Q}$ *and* $[P]$ *is the union of the two generators of* $\mathcal{Q}$ *passing through P, then* $\mathcal{Q}' = \mathcal{Q}\backslash[P]$ *gives rise to a code* $C_{Q'}$ *being a* $[16,4,9]_4$*-code admitting a PD-set* $\mathcal{S}$ *of size* 12.

Note that in case (b_1), the Gordon bound is 6.

4. The Klein quadric and the Schubert variety of $PG(5,q)$

Let $\mathfrak{U}_l$ be the set of all l-dimensional subspaces of $PG(r,q)$. The *Grassmann mapping* $\mathfrak{G} : \mathfrak{U}_l \to PG(N,q), N = \binom{r+1}{l+1} - 1$, associates to any $\mathcal{U} \in \mathfrak{U}_l$ a point $\mathfrak{G}(\mathcal{U})$ of $PG(N,q)$. Then $im\,\mathfrak{G} = \mathcal{G}_{l,r} := \mathfrak{G}(\mathfrak{U}_l)$ is an algebraic variety called the *Grassmannian* of the l-dimensional subspaces of $PG(r,q)$ (cf. [16], p. 107). The number of points of $\mathcal{G}_{l,r}$ is given by

$$|\mathcal{G}_{l,r}| = \begin{bmatrix} r+1 \\ l+1 \end{bmatrix} = \frac{(q^{r+1}-1)(q^r-1)\ldots(q^{r-l+1}-1)}{(q^{l+1}-1)(q^l-1)\ldots(q-1)}. \tag{7}$$

The Grassmannian $\mathcal{G}_{1,3}$ of the lines of $PG(3,q)$ is the hyperbolic quadric $\mathcal{KQ}$ of $PG(5,q)$ consisting of $(q^2+1)(q^2+q+1)$ points. It is called the *Klein quadric*.

It has a projective index 2, and it is covered by two systems of component planes. For general details see [14–16], for details on the intersection properties see [17] Section 4).

A linear code related to $\mathcal{KQ}$ is an $[n, k]$-code where $n = (q^2+1)(q^2+q+1), k = 6$ and minimum distance $d = q^4$ (see [18], p. 147, [13], p. 1579]).

Let $\mathcal{X} = \mathcal{X}_{\mathcal{KQ}}$ denote the projective system associated to $\mathcal{KQ}$. As usual, the basic parameters and spectrum are denoted respectively n, k, $d_t = d_t(\mathcal{X})$ and $A^{(s)}_{n-i} = A^{(s)}_{n-i}(\mathcal{X})$ with $s, t = 1{,}2{,}3{,}4$.

By direct computation we get

Proposition 5. $\mathcal{X}$ *has the following basic parameters and spectrum:*

(1) $n = (q^2+1)(q^2+q+1), k = 6$,

$$d_1 = q^4, d_2 = q^4 + q^3, d_3 = q^4 + q^3 + q^2, d_4 = q^4 + q^3 + 2q^2; \tag{8}$$

(2) *with respect to the* hyperplanes:

$A^{(1)}_{n-(q^3+2q^2+q+1)} = q^4 + q^3 + 2q^2 + q + 1$ *(hyperplanes cutting Schubert varieties);*

$A^{(1)}_{n-(q^3+q^2+q+1)} = q^5 - q^2$ *(hyperplanes cutting parabolic quadrics of the 4th dimension);*

and $A^{(1)}_{n-j} = 0,\ j \neq q^3 + q^2 + q + 1\ \ or\, j \neq q^3 + 2q^2 + q + 1.$

(3) *with respect to the* solids:

$A^{(2)}_{n-(2q^2+q+1)} = (q^3 + q^2 + q + 1)(q^2 + q + 1)$ *(solids cutting pairs of component planes);*

$A^{(2)}_{n-(q+1)^2} = \frac{1}{2}q^4(q^4 + q^3 + 2q^2 + q + 1)$ *(solids cutting hyperbolic quadrics of the 3th dimension);*

$A^{(2)}_{n-(q^2+q+1)} = (q^3 - q)(q^2 + 1)(q^2 + q + 1)$ *(solids cutting cones of the 3th dimension);*

$A^{(2)}_{n-(q^2+1)} = \frac{1}{2}q^4(q^3 - 1)(q - 1)$ *(solids cutting elliptic quadrics of the 3th dimension);*

and $A^{(2)}_{n-j} = 0,\ j \neq (q+1)^2, (q^2+1), (q^2+q+1), (2q^2+q+1).$

(4) *with respect to the* planes:

$A^{(3)}_{n-(q^2+q+1)} = 2(q^3 + q^2 + q + 1)$ *(component planes);*

$A^{(3)}_{n-(2q+1)} = \frac{1}{2}q^2(q+1)^2(q^2+1)(q^2+q+1)$ *(planes cutting two component lines);*

$A^{(3)}_{n-(q+1)} = (q^4(q^3-1)(q^2+1)) + ((q^4-1)(q^2+q+1)) = c + r$ *(planes cutting one conic (c conics) or one line (r lines));*

$A^{(3)}_{(n-1)} = \frac{1}{2}q^2((q^4+1)(q^2-q+1) - 2q^3)$ *(planes cutting one point);*

and $A^{(3)}_{(n-j)} = 0,\ j \neq (q^2+q+1), (2q+1), (q+1), 1$ *(there are no s-secant planes for* $s \in \{2, 3, \ldots, q\}$*)..*

(5) *with respect to the* lines:

$A^{(4)}_{n-(q+1)} = (q^2+1)(q^2+q+1)(q+1)$ *(component lines);*

$A^{(4)}_{n-2} = \frac{1}{2}q^4(q^4+q^3+2q^2+q+1)$ *(2-secant lines);*

$A^{(4)}_{n-1} = (q^3-q)(q^2+1)(q^2+q+1)$ *(tangent lines);*

$A^{(4)}_{n-0} = \frac{1}{2}q^4(q^3-1)(q-1)$ *(external lines);*

and $A^{(4)}_{n-j} = 0,\ j \neq (q+1),\ 2,\ 1,\ 0.$

From Section 2, it is clear that the previous Proposition 5 provides the complete spectrum of a linear code $\mathcal{C}_{\mathcal{X}}$ related to the Klein quadric.

In Section 5 of [19] is shown the following example.

Example 1 A binary linear code $\mathcal{C}_{\mathcal{KQ}}$ related to the Klein quadric $\mathcal{KQ}$ in $PG(5,2)$ is a [35, 6]-code with minimum distance $d = 2^4 = 16$ and admits a PD-set of size 40.

A *Schubert variety* $\mathcal{S}_{\mathcal{KQ}}$ of $PG(5,q)$ is a section of the Klein quadric $\mathcal{KQ}$ by a tangent hyperplane T. Thus $\mathcal{S}_{\mathcal{KQ}}$ is a cone of $2(q+1)$ planes with vertex $O \in \mathcal{KQ}$ and consists of $n = q^3 + 2q^2 + q + 1$ points. It corresponds via the Grassmann mapping $\mathfrak{G}$ to a *special linear complex of lines* of $PG(3,q)$, that is, a set comprising all lines meeting a fixed line. It is unique up to projectivities (cf. [14–16]).

If $(x_0, \ldots, x_5)$ are projective coordinates in $PG(5,q)$, the quadric $\mathcal{KQ}$ can be represented by the equation $x_0x_5 - x_1x_4 + x_2x_3 = 0$ so that a Schubert variety $\mathcal{S}_{\mathcal{KQ}}$ is the section of $\mathcal{KQ}$ by the hyperplane T with the equation $x_5 = 0$. Then $\mathcal{S}_{\mathcal{KQ}} = \mathcal{KQ} \cap T$ is a cone section with vertex the point (1, 0, 0, 0, 0) from which the hyperbolic quadric $\mathcal{H}_3 : x_1x_4 - x_2x_3 = 0$ of a solid is projected.

To calculate d_1 it is easy to verify that the maximum intersection with hyperplanes H of $\mathfrak{H}$ is obtained when H contains one of the planes and meets each of the remaining

q planes in lines, all passing through the vertex. Hence the maximum intersection is $2q^2 + q + 1$.

To calculate d_2 the maximum intersection with planes (2-dimensional subspaces of $\mathfrak{T}$) is obtained if a plane is one of the ruling planes or is a plane through the vertex which meets every ruling plane in a line. In any case, we get $q^2 + q + 1$.

Finally d_3 is clear as in $\mathcal{S}_{\mathcal{KQ}}$ there are component lines.

From above is proved the following result.

Proposition 6. *The basic parameters and weights of $\mathcal{S}_{\mathcal{KQ}}$ are*

$$n = q^3 + 2q^2 + q + 1,\ \ k = 5,\ \ d_1 = q^3 = n - 2q^2 + q + 1, \tag{9}$$

$$d_2 = n - q^2 + q + 1 = q^3 + q^2,\ \ d_3 = n - (q + 1) = q^3 + 2q^2. \tag{10}$$

If H is a hyperplane of $\mathfrak{H}$, then $|\mathcal{S}_{\mathcal{KQ}} \cap H| = (q + 1)^2$, or $|\mathcal{S}_{\mathcal{KQ}} \cap H| = 2q^2 + q + 1$ according to $O \notin H$ or $O \in H$, respectively. Therefore linear codes related to $\mathcal{S}_{\mathcal{KQ}}$ and to $\mathcal{V} = \mathcal{S}_{\mathcal{KQ}} \backslash \{O\}$ (both of dimension 5) have the same minimum distance $d = q^3$. Since the automorphisms group Aut $\mathcal{S}_{\mathcal{KQ}}$ fixes the vertex O, *a code related to $\mathcal{V}$ has better parameters than a code related to $\mathcal{S}_{\mathcal{KQ}}$.*

In Section 5 of [19] is shown the following example.

Example 2. A binary linear code $\mathcal{C}_{\mathcal{V}}$ related to the Schubert variety $\mathcal{V} = \mathcal{S}_{\mathcal{KQ}} \backslash \{O\}$ in $PG(5, 2)$ is a [18, 5]-code with minimum distance $d = 2^3 = 8$ and admits a PD-set of size 9.

To get some information about the Grassmannian $\mathcal{G}_{l,r}$ of the l-dimensional subspaces in $PG(r, q), r > 5$, and the Schubert variety $\Omega(\alpha) \subset \mathcal{G}_{l,r}$ (where $\alpha = (a_0, \ldots, a_l)$ is the corresponding sequence of dimensions) and their codes, see Theorem 12 of [9, 20–22].

5. The rational ruled surfaces V_2^{r-1} of $PG(r, q)$

Let us consider varieties of P^r with $u = 2$ and $v = r - 1$.

The following result is well known (see [15]).

Proposition 7 *The varieties V_2^{r-1} of P^r are the rational ruled varieties and the Veronese surface if $r = 5$.*

Suitably modified it can be easily proved also for the finite case.

Assume $r \neq 5$. Denote S_t a projective t-dimensional subspace of P^r for $t < r$.

Proposition 8

i. *V_2^{r-1} is a ruled rational normal surface.*

ii. *A variety V_2^{r-1} contains irreducible rational normal curves C^t of order $t \leq r - 1$ each of them existing in a t-dimensional subspace S_t.*

iii. *Each V_2^{r-1} is a ruled surface obtained by means of a projectivity between two irreducible directrix curves C^m and C^{r-m-1} contained in an m-dimensional subspace S_m and in an $(r - m - 1)$-dimensional subspace S_{r-m-1}, respectively.*

iv. *Let m be the order of the minimum order directrix of V_2^{r-1}. Then h generatrix lines are dependent if $h \leq m+1$, otherwise they are independent.*

v. *If $r = 2s$, then $m = (r-2)/2$ is the order of the minimum order directrix, if $r = 2s+1$, then there exist directrix curves of order $m \leq (r-1)/2$.*

Choose and fix a surface V_2^{r-1} with a minimum order directrix C^m with $m < q$.

Denote $\mathcal{X}$ the projective system of the rational points of V_2^{r-1}, C the linear code related to $\mathcal{X}$. It is $|\mathcal{X}| = (q+1)^2$.

Proposition 9 *C is an $[n,k,d]_q$-code with*

$$n = (q+1)^2, k = r+1, d = d_1 = q^2 - mq, d_{r-1} = q^2 + q. \qquad (11)$$

For the proof see [20], Theorem 4.

From $d_1 \leq n - k + 1$, the definition of genus of a code and from Proposition 9 follows

Proposition 10 (1) *The inequality $(m+2)q \geq r-1$ holds for every q and r.*
(2) *C is of genus at most $g \geq (m+2)q - (r-1)$*

Consider now the case $r = 4$. Denote $\mathfrak{H}$ the set of the hyperplanes. Let V_2^3 be a ruled surface of $P^4 = PG(4,q)$. We have named V_2^3 the *celtic variety* for its *hut shape* (see [4], Section 4).

From Proposition 8, from Lemma 7 of [20] and Propositions 1.1, 1.3, 1.4, 1.5, and Theorem 1.2 of [23] we obtain

Lemma 11

a. *A celtic variety V_2^3 is constructed by means of a projectivity connecting the points of the minimum order directrix, the line l, with the points of a non-degenerate conic C^2 of a plane π with $\pi \cap l = \emptyset$.*

b. *V_2^3 has $q+1$ two by two skew generatrix lines. They connect birationally the points of l and the points of C^2.*

c. *There exists a unique hyperplane H such that $l \subset H$ and $l' = \pi \cap H$ is skew to l.*

d. *There exist hyperplanes H' such that one generatrix line g_1 belongs to H' and $H' \cap V_2^3 = \left\{g_1, C'^2\right\}$ for some conic C'^2.*

e. *For every two generatrix lines $g_i, g_j, i \neq j$ the hyperplane $H' = g_i \cup g_j$ is such that $H' \cap V_2^3 = \left\{g_i, g_j, l\right\}$. Such hyperplanes have the maximum intersection with V_2^3.*

f. *No hyperplane contains 3 generatrix lines.*

g. *Every two points $P, Q \in V_2^3$ belong to l, or to a generatrix line g, or to a unique conic of V_2^3.*

h. *A plane π' can meet V_2^3 either in one point, or in one line, or in one irreducible conic, or it is the intersection of l with e generatrix line g. A plane $\pi' = l \cup g$ is a tangent plane and $|\pi' \cap V_2^3|$ is maximum.*

i. *The totality of varieties V_2^3 of $PG(4,q)$ having l and C^2 as directrices are projectively equivalent and their number is $(q+1)q(q-1)$.*

Denote $\mathcal{X}$ the projective system consisting of the rational points of V_2^3 and $C_{\mathcal{X}}$ a linear code associated to it.

From Proposition 9, Proposition 10, 2), Lemma 11, (*a*), (*h*) and from [20], Theorems 8 and 9, we obtain

Proposition 12 *$C_{\mathcal{X}}$ is an $[n,k]_q$-code with $n=(q+1)^2, k=5, d_1=q^2-q, d_2=q^2, d_3=q^2+q. C_{\mathcal{X}}$ (and $C_{\mathcal{X}}^{\perp}$) is of positive genus $g \geq 3q-3$.*

The spectrum $A_i^{(1)}$ of $\mathcal{X}$ is

$$A_{d_1}^{(1)}=(q+1)\frac{q}{2},\quad A_{d_2}^{(1)}=(q^2-q)(q+1),\quad A_{d_3}^{(1)}=(q^4+1)+\frac{q(q+3)}{2}, \tag{12}$$

$$A_i^{(1)}=0 \quad for \quad all \quad i\in\{1,2,\dots,n\}\backslash\{d_1,d_2,d_3\}. \tag{13}$$

Denote g_s the generatrix line joining corresponding points $L_s \in l$ and $C_s \in C$. As l is the unique line intersecting all generatrices and there are no other lines contained in $\mathcal{X}$ than l and the generatrix, follows that every automorphism $\alpha \in \text{Aut } \mathcal{X}$ fixes the directrix line l and maps every generatrix g_s to a generatrix g_s' (cf. [4], Lemma 3).

From Lemma 11 follows that the intersection of $\mathcal{X}$ with a hyperplane H is the union of a generatrix and a conic, or the union of two generatrices and l, or the union of one generatrix and l, or l, or a cubic curve. Hence $\max\{|\mathcal{X}\cap H| \; |H\in\mathfrak{H}\}=3q+1$.

In order to construct PD-sets for the related codes, in Proposition 14 of [4], two subgroups of Aut $\mathcal{X}$, namely $\mathcal{A}$ and $\mathcal{N}$, are chosen. $\mathcal{A}$ is isomorphic to the group of the affine bijections of F: $\{x|x\mapsto xm+b, m,b\in F, m\neq 0\}$, $\mathcal{N}$ fixing each generatrix line is a normal subgroup.

Let $\mathcal{X}'=\mathcal{X}\backslash\{l\}$. Note that $|\mathcal{X}'|=q^2+q$ and $\max\{|\mathcal{X}'\cap H| \; |H\in\mathfrak{H}\}=3q+1-(q+1)=2q$ so that the codes $C_{\mathcal{X}}$ and $C_{\mathcal{X}'}$ have the same minimum distance.

As $\mathcal{X}$ generates $PG(4,q)$, choose a subset $\mathcal{I}\subset\mathcal{X}$ of independent points.

From above and by comparing [4], Proposition 15, we get the following result.

Proposition 13 (1) *$C_{\mathcal{X}}$ is a $\left[(q+1)^2,5,q(q-1)\right]_q$-code.*

(2) *$C_{\mathcal{X}'}$ is a $[q(q+1),5,q(q-1)]_q$-code.*

(3) *If $q\geq 4, \mathcal{I}\subset\mathcal{X}$ an independent set of $PG(4,q)$ with $\mathcal{I}\cap l\neq\emptyset$, then there is no PD-set for $\mathcal{I}$.*

From [4], Propositions 17 and 19 follows

Proposition 14 (1) *If $\mathcal{X}$ is in $PG(4,3)$, then $C_{\mathcal{X}}$ is a 2-error-correcting $[16,5,6]_3$-code admitting a PD-set S of minimum size 3.*

(2) *If $\mathcal{X}$ is in $PG(4,4)$ and $\mathcal{X}'=\mathcal{X}\backslash l$, then the code $C_{\mathcal{X}'}$ is a 5-error-correcting $[20,5,12]_4$-code admitting a PD-set S of size 24.*

6. Ruled sets

Let $PG(k-1,q)=(\mathcal{P},\mathfrak{L})$ be a $(k-1)$-dimensional projective space over $F=GF(q), k\geq 3$ with point set $\mathcal{P}$ and line set $\mathfrak{L}$. Denote $\mathfrak{H}$ the set of the hyperplanes of $PG(k-1,q)$.

Let $\mathcal{K} \subset \mathcal{P}$. Denote $M_{\mathcal{P}}$ the *type of* $\mathcal{K}$ *with respect to hyperplanes* (that is, the set of all intersection numbers of $\mathcal{K}$). For $i \in M_{\mathcal{P}}$ let $t_i := \{H \in \mathfrak{H} | \ |\mathcal{K} \cap H| = i\}$ denote the total number of hyperplanes yielding the intersection number i.

If $\mathcal{K}$ is a projective system of type $M_{\mathcal{P}}$, then for $i \in M_{\mathcal{P}}$ there are t_i code-words in $C_{\mathcal{K}}$ of weight $|\mathcal{K}| - i$.

From Lemma 1 of [24] we obtain

Lemma 15 *Let* $\mathcal{S} \subset \mathcal{P}$ *be a subspace with* $\emptyset \neq \mathcal{S} \neq \mathcal{P}$ *and* $\mathcal{K} \subset \mathcal{S}$. *Then* $M_{\mathcal{P}} = M_{\mathcal{S}} \cup \{|\mathcal{K}|\}$.

Let $\mathcal{S}$ and $\mathcal{S}'$ be two complementary subspaces in $PG(k-1, q)$. Choose and fix two subsets $\mathcal{K} \subset \mathcal{S}$ and $\mathcal{K}' \subset \mathcal{S}'$ with $\overline{\mathcal{K}} = \mathcal{S}, \overline{\mathcal{K}'} = \mathcal{S}'$. Set $m := |\mathcal{K}|, m' := |\mathcal{K}'|$.

Denote $\mathcal{R} = \{\overline{x, x'} | x \in \mathcal{K},\ x' \in \mathcal{K}'\}$. A *ruled set* $\mathcal{X}$ is the set of the points of the lines of $\mathcal{R}$, that is, $\mathcal{X} := \bigcup_{X \in \mathcal{R}} X$.

From [24], Lemmas 2 and 3 follows

Lemma 16

1. *Let* $x_1, x_2 \in \mathcal{K}$ *and* $x'_1, x'_2 \in \mathcal{K}'$ *with* $x_1 \neq x_2,\ \ x'_1 \neq x'_2$; *then* $\overline{x_1, x'_1} \cap \overline{x_2, x'_2} = \emptyset$.

2. *Let* $L_1, L_2 \in \mathcal{R}, L_1 \neq L_2$ *with* $L_1 \cap L_2 \neq \emptyset$; *then* $L_1 \cap L_2 \in \mathcal{K} \cup \mathcal{K}'$.

3. $|\mathcal{R}| = m\, m'$.

4. $|\mathcal{X}| = m\, m'(q-1) + m + m'$.

5. *If* $H \in \mathfrak{H}$ *is a hyperplane and* $m_H := |H \cap \mathcal{K}|,\ \ m'_H := |H \cap \mathcal{K}'|$, *then* $|H \cap \mathcal{X}| = m_H \cdot m'_H(q-1) + m_H + m'_H + (m - m_H) \cdot (m' - m'_H)$.

6. *The linear code* $C_{\mathcal{X}}$ *has length* $|\mathcal{X}| = m\, m'(q-1) + m + m'$ *and dimension k.*

If $\mathcal{K} = \mathcal{S}$ and $\mathcal{K}' = \mathcal{S}'$ then $\mathcal{X} = \mathcal{P}$, hence $C_{\mathcal{X}} = C_{\mathcal{P}}$ is the simplex code of dimension k. As each hyperplane H is contained in $\mathcal{X}$, in such a case every code word has the same weight. Therefore the minimum distance (or, weight) is $d = |\mathcal{X}| - |H| = q^{k-1}$.

From [24], Lemma 4 follows

Result If $H_{\mathcal{S}} \subset \mathcal{S}$ and $H_{\mathcal{S}'} \subset \mathcal{S}'$ are subspaces of $\dim H_{\mathcal{S}} = \dim \mathcal{S} - 1$ and $\dim H_{\mathcal{S}'} = \dim \mathcal{S}' - 1$, then there exist exactly $q+1$ hyperplanes H with $H_{\mathcal{S}}, H_{\mathcal{S}'} \subset H$ one of which contains $\mathcal{S}$ and one contains $\mathcal{S}'$.

Let $M_{\mathcal{S}}$ and $M_{\mathcal{S}'}$ be the type of $\mathcal{K}$ and $\mathcal{K}'$ with respect to hyperplanes, respectively. Denote $M = M_{\mathcal{S}} \cup \{m\}, M' = M_{\mathcal{S}'} \cup \{m'\}, m_o = \min M, m'_o = \min M', m_1 = \max M_{\mathcal{S}}$ and $m'_1 = \max M_{\mathcal{S}'}$.

Consider the following mapping

$$\iota : M \times M \to \mathbf{N},\quad \iota(a, a') = a'(aq + 1 - m) + a(1 - m') + m \cdot m'. \tag{14}$$

Then the type of $\mathcal{X}$ is $M_{\mathcal{X}} = \{\iota(a, a') | (a, a') \in M \times M \backslash \{(m, m')\}\}$ and $\max M_{\mathcal{X}} = \max\{\iota(m_o, m'_o),\ \iota(m, m'_1),\ \iota(m_1, m')\}$ (cf. [24], Proposition 5 and Lemma 6).

Hence we can determine the weight distribution of $C_{\mathcal{X}}$ once known the types $M_{\mathcal{S}}$ and $M_{\mathcal{S}'}$ of $\mathcal{K}$ and $\mathcal{K}'$, respectively.

Proposition 17 *The code* $C_{\mathcal{X}}$ *is a linear code of length* $n = \iota(m, m')$, *dimension k and minimum weight* $d = n - \max\{\iota(m_o, m'_o),\ \iota(m, m'_1),\ \iota(m_1, m')\}$.

See [24] Theorem 7.

Since $\mathcal{X}$ generates the projective space $(\mathcal{P}, \mathfrak{L})$ there exists a basis $\mathcal{B} \subset \mathcal{X}$ of $(\mathcal{P}, \mathfrak{L})$. Let $\mathcal{X} = \{p_1, \dots, p_n\}$ such that $\mathcal{B} = \{p_1, \dots, p_k\}$ is a basis. Let $\mathbf{v}(\mathcal{X})$ be a system of vectors representing $\mathcal{X}$. For $p_j \in \mathcal{X}$ let $\mathbf{v}(p_j) = \sum_{i=1}^{k} \gamma_{ij} \mathbf{v}(p_i)$ be the vector representing the point p_j with respect to the basis $\mathbf{v}(\mathcal{B})$ of F^k. Then $G = (\gamma_{ij})$ is a standard generatrix matrix of the code $C_{\mathcal{X}}$.

If $\mathcal{B}_{\mathcal{K}} \subset \mathcal{K}$ is a basis of $\mathcal{S}$ and $\mathcal{B}_{\mathcal{K}'} \subset \mathcal{K}'$ is a basis of $\mathcal{S}'$ then $\mathcal{B} = \mathcal{B}_{\mathcal{K}} \cup \mathcal{B}_{\mathcal{K}'} \subset \mathcal{X}$ is a basis of $(\mathcal{P}, \mathfrak{L})$. With such a basis it is easy to write down the standard generatrix matrix, in particular in the binary case.

For $q = 2$, the standard generatrix matrix $\mathcal{G}$ for $C_{\mathcal{X}}$ is shown in [24], p.751.

For the following Examples see Examples 1, 2 and 3 of [24], pp. 751–754.

Example 3 In $PG(k-1, q), k \geq 5$, choose and fix an ellipsoid $\mathcal{E}$ in a 3-dimensional subspace $\mathcal{S}$, let $\mathcal{S}'$ be a complementary subspace of $\mathcal{S}$, set $r := \dim \mathcal{S}' = k - 5$.

The code $C_{\mathcal{X}}$ associated to the ruled set defined by $\mathcal{K} = \mathcal{E}, \mathcal{K}' = \mathcal{S}'$ has length $n = (q^2+1)\left(\sum_{i=0}^{r} q^i\right)(q-1) + q^2 + 1 + \sum_{i=0}^{r} q^i = q^{r+3} + \sum_{i=0}^{r+1} q^i$ and dimension $k = r + 5$.

The type of $\mathcal{K} = \mathcal{E}$ is $M_{\mathcal{S}} = \{m_o = 1, m_1 = q+1\}$; it holds $t_{m_o} = q^2 + 1$ and $t_{m_1} = q^3 + q$.

The type of $\mathcal{K}' = \mathcal{S}'$ is $M_{\mathcal{S}'} = \left\{m'_0 = \sum_{i=0}^{r-1} q^i\right\}$ and $t_{m'_o} = \sum_{i=0}^{r} q^i$.

Then the weight distribution is obtained. There are:

- $q^2 + 1$ code words of weight q^{r+3},
- $\sum_{i=4}^{r+4} q^i$ code words of weight $q^{r+3} - q^{r+2} + q^{r+1}$,
- $q^3 + q$ code words of weight $q^{r+3} - q^{r+2}$.

This shows that the minimum weight of $C_{\mathcal{X}}$ is $d = q^{r+3} - q^{r+2}$.

For $q = 2$ the code $C_{\mathcal{X}}$ is a linear $\left[2^{r+3} + 2^{r+2} - 1, r+5, 2^{r+2}\right]$-code with error-correcting capability $t = 2^{r+1} - 1$.

For $r = 1$ the code $C_{\mathcal{X}}$ is a [23, 6, 8]-code.

Example 4 In $PG(5, q), q = 2^h$, choose and fix two ovals with their nucleus, $\mathcal{K} \subset \mathcal{S}$ and $\mathcal{K}' \subset \mathcal{S}'$, respectively, where $\mathcal{S}$ and $\mathcal{S}'$ are two skew planes.

The code $C_{\mathcal{X}}$ associated to the ruled set defined by $\mathcal{K}$ and $\mathcal{K}'$ has length $n = (q+2)(q+2)(q-1) + 2(q+2) = q^3 + 3q^2 + 2q$ and dimension $k = 6$.

There are $\binom{q+2}{2} = \frac{1}{2}(q^2 + 3q + 2)$ lines in $\mathcal{S}$ and $\mathcal{S}'$ meeting $\mathcal{K}$ and $\mathcal{K}'$ in 2 points and $\frac{1}{2}(q^2 - q)$ lines in $\mathcal{S}$ and $\mathcal{S}'$ missing $\mathcal{K}$ and $\mathcal{K}'$, respectively.

We obtain the following weight distribution. There are

- $q^2 - q$ code words of weight $q^3 + 3q^2 + q - 2$,
- $\frac{1}{2}(q^5 + q^4 - 3q^3 - q^2 + 2q)$ code words of weight $q^3 + 2q^2 - 2$,
- $\frac{1}{4}(q^5 + 5q^4 + 7q^3 - q^2 - 8q - 4)$ code words of weight $q^3 + 2q^2 - 2q$,
- $\frac{1}{4}(q^5 - 3q^4 + 3q^3 - q^2)$ code words of weight $q^3 + 2q^2 - 2q - 4$,

- $q^2 + 3q + 2$ code words of weight $q^3 + q^2 - q$.

This shows that the minimum weight of $C_\mathcal{X}$ is $d = 8$ for $q = 2$ and $d = q^3 + q^2 - q$ for $q \geq 4$.

For $q = 2$ the code $C_\mathcal{X}$ is a linear [24, 6, 8] -code with error-correcting capability $t = 3$.

Example 5 In $PG(7, q)$ choose and fix two ellipsoids $\mathcal{E}$ and $\mathcal{E}'$ in two non-intersecting 3-dimensional subspaces $\mathcal{S}$ and $\mathcal{S}'$, respectively. Then the code $C_\mathcal{X}$ related to the ruled set defined by $\mathcal{K} = \mathcal{E}, \mathcal{K}' = \mathcal{E}'$ has length $n = (q^2 + 1)(q^2 + 1)(q - 1) + 2q^2 + 2 = q^5 - q^4 + 2q^3 + q + 1$ and dimension $k = 8$.

There are $q^2 + 1$ planes $E \subset \mathcal{S}$ with $|E \cap \mathcal{E}| = 1$ and $q^3 + q$ planes $E \subset \mathcal{S}$ with $|E \cap \mathcal{E}| = q + 1$.

We obtain the following weight distribution.

There are

- $2(q^2 + 1)$ code words of weight $q^5 - q^4 + q^3$,
- $2\left(q^6 - q^5 + 2q^4 - 2q^3 + q^2 - q\right)$ code words of weight $q^5 - 2q^4 + 3q^3 - q^2$,
- $q^7 - q^6 + 2q^5 - 2q^4 + q^3 - q^2$ code words of weight $q^5 - 2q^4 + 3q^3 - 2q^2$,
- $q^5 - q^4 + 2q^3 - 2q^2 + q - 1$ code words of weight $q^5 - 2q^4 + 2q^3$,
- $2q(q^2 + 1)$ code words of weight $q^5 - 2q^4 + 2q^3 - q^2$.

This shows that the minimum weight of $C_\mathcal{X}$ is $d = q^5 - 2q^4 + 2q^3 - q^2$.

For $q = 2$ the code $C_\mathcal{X}$ is a linear [35, 8, 12]-code with error-correcting capability $t = 5$.

In [24], pp. 752–753 the standard generatrix matrices of the three examples are shown.

7. Conclusions

The close connection between the geometry of the projective varieties, or in general, of suitable subsets of a finite geometry and linear codes through projective systems, certainly still has prospects for interesting developments. This is, on the one hand, because of the elaboration and study of eventually new varieties, and, on the other, for the possibility of constructing linear codes with interesting parameters for the various applications in the communication systems.

Acknowledgements

Dedicated to Professor Hans-Joachim Kroll for his 80th birthday.

Classification:

Mathematics Subject Classification (2020): 94B05, 94B27, 51E20, 51A22

Author details

Rita Vincenti
Department of Mathematics and Computer Science, University of Perugia, Perugia, Italy

Address all correspondence to: aliceiw213@gmail.com

References

[1] Macwilliams FJ. Permutation decoding of systematic codes. Bell System Technology Journal. 1964;**43**: 485-505

[2] Gordon DM. Minimal permutations sets for decoding the binary Golay codes. IEEE Transactions on Information Theory. 1982;**IT-82**:541-543

[3] Huffman WC. Codes and groups. In: Pless VS, Huffman WC, editors. Handbook of Coding Theory. Elsevier Science B.V; 1998:1345-1440

[4] Kroll H-J, Vincenti R. PD-sets for the codes related to some classical varieties. Discrete Mathematics. 2005;**301**:89-105

[5] Kroll H-J, Vincenti R. Antiblocking systems and PD-sets. Discrete Mathematics. 2008;**308**:401-407

[6] Kroll H-J, Vincenti R. Antiblocking decoding. Discrete Applied Mathematics. 2010;**158**:1461-1464

[7] Wei V. Generalized hamming weights for linear codes. IEEE Transactions on Information Theory. 1991;**37**:1412-1418

[8] Ryan CT. An application of Grassmann varieties to coding theory. Congressus Numerantium. 1987;**57**: 257-271

[9] Ghorpade SR, Lachaud G. Higher weights of Grassmann codes. In: Proc. Int. Conf. on Coding Theory, Cryptography and Related Areas (Guanajuato, Mexico, 1998). Berlin/Heidelberg: Springer-Verlag; 2000:122-131

[10] Montanucci E, Vincenti R. Characterization of Projective Systems Related to Linear Codes. Italy: University of Perugia: Tech. Report DMI 2003/09

[11] Pless VS, Huffman WC, editors. Handbook of Coding Theory. Amsterdam: Elsevier; 1998

[12] Tsfasman MA, Vladut SG. Algebraic Geometric Codes. Amsterdam: Kluwer; 1991

[13] Tsfasman MA, Vladut SG. Geometric approach to higher weights. IEEE Transactions on Information Theory. November 1995;**41**(6): 1564–1588

[14] Hirschfeld JWP. Finite Projective Spaces of Three Dimensions. Oxford: Clarendon Press; 1985

[15] Bertini E. Introduzione alla geometria proiettiva degli iperspazi. Pisa: Enrico Spoerri Editore; 1907

[16] Hirschfeld JWP, Thas JA. General Galois Geometry. Oxford: Oxford University Press; 1991

[17] Kroll H-J, Vincenti R. Partitions of the Klein quadric, proceedings of Combinatorics'06, Ischia (Naples), June 25-July 1st, 2006. Electronic Notes in Discrete Mathematics Volume. 2006; **26**(1):151-157

[18] Yu D. NOGIN, Codes associated to Grassmannians, Arithmetic Geometry and Coding Theory 4. Berlin/New York: W.de Gruyter and Co; 6:145-154

[19] Kroll H-J, Vincenti R. PD-sets for binary RM-codes and the codes related to the Klein quadric and to the Schubert variety of PG(5, 2). Discrete Mathematics. 2008;**308**:408-414

[20] Vincenti R. On Some Classical Varieties and Codes. Italy: University of Perugia; Tech. Report DMI 2000/20

[21] Ghorpade SR, Tsfasman MA. Basic Parameters of Schubert Codes. 2002

[22] Guerra L, Vincenti R. On the Linear Codes Arising from Schubert Varieties, Designs, Codes and Cryptography. Vol. 33. The Netherlands: Kluwer Academic Publisher; 2004:173-180

[23] Bernasconi C, Vincenti R. Spreads induced by varieties V_2^3 of $PG(4, q)$ and Baer subplanes. BUMI. 1981;**18**:197-207

[24] Kroll H-J, Vincenti R. Linear codes from ruled sets in finite projective spaces. Designs Codes and Cryptography (DESI). 2020;**88**:747-754

Chapter 4

Hybrid Obfuscation of Encryption

Asma'a Al-Hakimi and Abu Bakar Md Sultan

Abstract

Obfuscation is an encryption method. It allows the programmer to reform the code for protection. Obfuscation has promising chance to change the way of coding, where the programmer has the ability to program with any language, not necessarily English. Obfuscation, Unicode, and mathematical equations have the possibility to change strings and identifiers and to hide secret algorithms and business rules. Special dictionary is used for the string obfuscation to hide the logic of the program. The hybrid obfuscation technique will be implemented into a tool that automatically converts the code. It can be can be used for games and mobile applications for protection. With obfuscation, the application still has the ability to perform sufficiently and provide the desired output without any delays in performing timing. After obfuscating the source file, reverser still has the ability to break the object file but will not be able to read or understand and when obfuscation technique is complicated, reversing leads to error where original code disappears. In this chapter, hybrid obfuscation will be presented with examples, and obfuscation table is presented as well for future use.

Keywords: obfuscation, encryption, anti-reverse engineering, reverse engineering, hacking prevention

1. Introduction

Obfuscation is considered as anti-reverse engineering to prevent hacking and code theft. It mainly works in the source file to change the form of the code to confuse the reverser or the hacker and also to prevent the compiler from reading the hacked code. The obfuscation technique converts all the code into unreadable text, but it functions like the original code and produces the same output. There are many forms of obfuscation, such as string encryption, hiding, changing identifier names, junk code obfuscation, packing, byte code obfuscation, string encryption, stealth obfuscation, chaotic encryption, and junk code obfuscation. In this chapter, a new method of obfuscation is introduced to produce a different kind of chaotic code that is almost impossible to read and understand but still produces the desired output [1]. For this case, Java code will be used to implement and test the code. **Figure 1** presents most common categories of obfuscation.

The decision of using any category of any obfuscation or merging them together depends on the level of complication the programmer or author wish to make the code and also depends on the part that wanted to be obfuscated, such as a business rule or an algorithm that is important to the code or the business of the company that is developing the code. Following section describes the categories of the obfuscation.

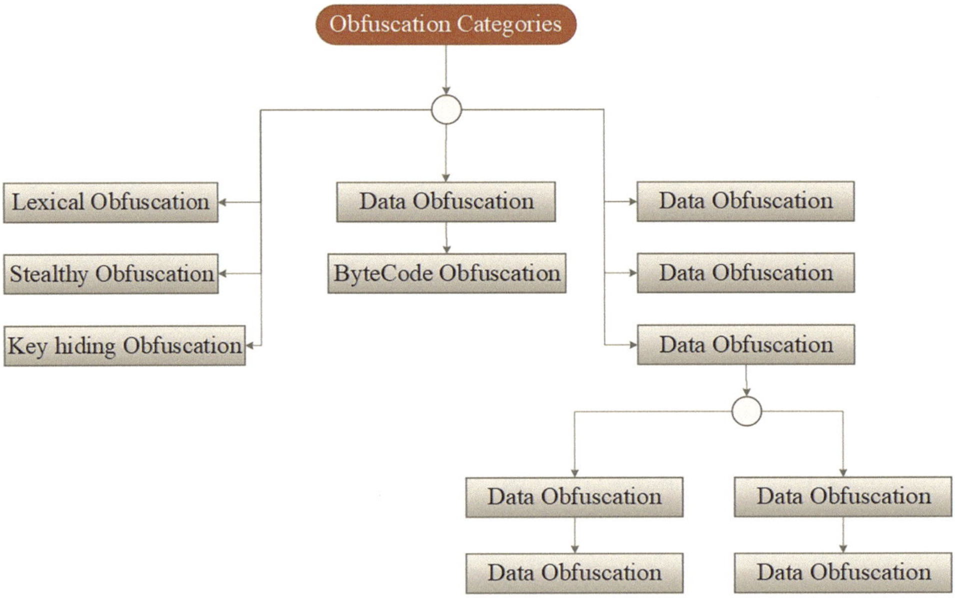

Figure 1.
Obfuscation categories.

1.1 Lexical obfuscation

This technique is used to transform or alter the compiler information. Other information will be removed from the byte code such as comments and identifiers. Programmers use this technique alone without merging it with any other technique that does not guarantee protection [2].

1.2 Stealthy obfuscation

This is an obfuscator that contains several obfuscation techniques to obfuscate the code when it is read. Stealthy obfuscation provides a sort of false sense of the actual program structure. This technique works with the assembly file. After applying this technique, two files are created, one of them is the assembly file and the second file is the obfuscated file. In this technique, the source file is not encrypted. However, there is a possibility that the reverser will be confused when reading the code [3].

1.3 Key hiding obfuscation

This technique is used to protect intellectual property, and it is based on key hiding. This method should not be used alone. It must be combined with another technique to provide more protection. A symmetric mechanism is used to combine with key hiding. Key hiding focuses on executable software. The software protection key is then encrypted with a threshold key to make it difficult for the reverser to find it and break the code. This technique focuses on executable software and leaves the source code and class file as they are [4].

1.4 Junk obfuscation

This technique converts identifiers into an unreadable but performs and produces output that can be read by the compiler. This technique is particularly useful when combined with another technique to increase code security level. Junk obfuscation misleads and does not allow the reverser to read the identifier or understand what is the purpose for it but can only see the output [5].

1.5 Control obfuscation

This technique hides the actual flow of the code and creates a fake one. It controls flow by using a structured exception handling mechanism in Windows. It disguises the control flow by adding exception statements. When the exception occurs, the exception handler is called and the flow of execution is changed in the exception handler. This technique is provided by Windows operating system for exception handling. It focuses on basic blocks that can be obfuscated by further splitting them into few parts. This technique does not modify the class file. It changes the source file, which is a good point to protect the code. However, newer reverse engineering tools can change the flow of the software and even create new flows [6].

1.6 String obfuscation

This technique uses many approaches such as encryption, mathematical equations, or chaotic obfuscation. It depends on the programmer to decide how complicated the obfuscated strings should be. String obfuscation is very effective in protecting the code from theft. When the string is obfuscated, only the compiler can read and output it, while it becomes unreadable to humans [7].

1.7 Chaotic obfuscation

Here, a mathematical modeling is used for string encoding. It is up to the programmer to determine the form of the equation for encrypting the string. The programmer has the option to encrypt all strings or some of them. Chaos theory involves stems generated from mathematical equations that produce random numbers and chaos that are not readable by the user; however, the chaos sequences are readable by the compiler at runtime. The chaotic equations are deterministic by nature, which means that they go into saturation after several iterations at a single value. **Figure 2** presents sample of string after applying chaotic obfuscation [8].

1.8 Cipher algorithm

This technique uses session keys instead of permeant. Session keys are symmetric keys that are regenerated for each encryption. The keys in Cipher are automatically generated in the algorithm itself to prevent the inverter from guessing the permanent key. The user using Cipher can purchase a permanent key from the developer; however, the key can be compromised by determined reversers. String encryption in Cipher follows certain steps: the first step is to choose the secret key, which can be an x-value. The second step is to assign the equation used for encryption that will to cause the series of chaos. The encryption is a secret function that only the developer

```
public static void main(String[] args)throws Exception

{byte[] input = new byte[] {
        0x00, 0x11, 0x22, 0x33, 0x44, 0x55, 0x66, 0x77,
        0x08, 0x09, 0x0a, 0x0b, 0x0c, 0x0d, 0x0e, 0x0f,
        0x10, 0x11, 0x12, 0x13, 0x14, 0x15, 0x16, 0x17};

byte[] keyBytes = new byte[]{
        0x00, 0x01, 0x02, 0x03, 0x04, 0x05, 0x06, 0x07,
        0x08, 0x09, 0x0a, 0x0b, 0x0c, 0x0d, 0x0e, 0x0f,
        0x10, 0x11, 0x12, 0x13, 0x14, 0x15, 0x16, 0x17};

SecretKeySpec key = new SecretKeySpec(keyBytes, "AES");

Cipher cipher = Cipher.getInstance("AES/ECB/PKCS7Padding", "BC");

System.out.println("input text : " + input);//Utils.toHex(input));
```

Figure 2.
String chaotic obfuscation.

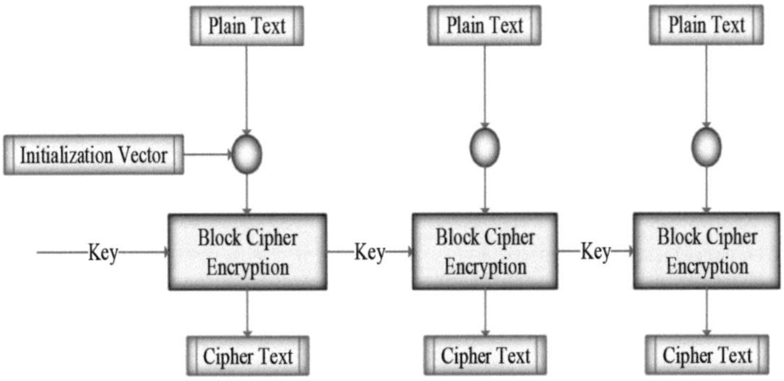

Figure 3.
Cipher obfuscation.

knows. The third step is the iteration of x-value to produce the ciphertext. **Figure 3** presents sample of cipher obfuscation [9].

1.8.1 Cipher block chaining

This technique divides the data or code into blocks of bits and chains. The encrypted data are blocked together to avoid eavesdroppers from inserting their own blocks of bits among the blocks of encrypted code. A mathematical equation is used for the for the Cipher block changing, and the equation is given as follows:

$$C_1 = \text{ek}(m_1\text{XORIV}) \tag{1}$$

$$C_1 = \text{ek}(m_1 \text{XOR} C_{i-1}) \quad \textit{for} \;\; \text{i} > 1 \tag{2}$$

The technique involves a specific (N) value passed between the plaintext to ensure that the ciphertext blocks look different. The N value is the second layer of encryption, while the first layer of encryption is done by the secret key. Each generated text is encrypted with the same secret key. If an error occurs in one of the blocks, it will also occur in all other blocks that follow the affected block [10].

1.9 Symmetric cipher

The symmetric Cipher is well known and common for string encryption and decryption. It can encrypt large data. This technique uses one key of encryption and decryption. The reverser or the end user must find the meaning of exchanging the key securely. Without the key of the encryption algorithm, the reverser will not be able to reveal or translate or decrypt the encrypted string. Below figure illustrates the sample of code before and after applying Cipher algorithm. **Figure 4** presents sample of symmetric obfuscation after applying on **Java code [11]**.

2. Discussion of current obfuscation techniques

Obfuscation techniques based on the identifiers renaming have been recently presented. Such techniques can be classified as a form of layout obfuscation, since they reduce the information available to a human reader which examines the target program, or of preventive obfuscation since they aim to prevent the decompilation from producing original code with full meaning or to produce an incorrect Java source code. Such techniques try to hide the structure and the behavior information embedded in the identifiers of a Java program by replacing them with meaningless or confounding identifiers to make more difficult the task of the reverse engineer. It is worth to notice that the information associated with an identifier is completely lost after the renaming [12]. By replacing the identifiers of a Java bytecode with new ones that are illegal with respect to the Java language specification, such techniques try to make the decompilation process impossible or make the decompiler return unusable source code. After applying any obfuscation technique, it is very important to test the program, especially if there are many loops that execute many times such as games or algorithmically intensive method. The constant test is to ensure that all obfuscated

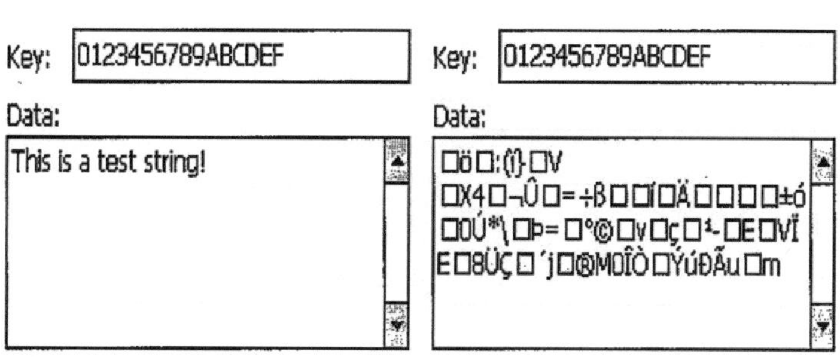

Figure 4.
Symmetric cipher obfuscation.

parts are well working with no error recorded. During test for several obfuscation techniques, there were several limitations that can be a vulnerable entrance for the any strong decompiler [13].

Control flow obfuscation is only able to defeat decompilers when the method contains basic blocks of code. This technique is not fully deterministic, whereby it is only applicable to methods if the developer sees the performance degradation during testing. If the control flow obfuscation was implemented on highly complicated code that contains extensive loops, it will not be useful as it will be difficult to trace the errors during implementation, and it does not work sufficiently during reversing. However, it is useful with small applications. Control flow obfuscation does not have the ability to be nested in the source file, as it will be difficult to trace the loops during execution. Error management will not be as possible as it should be [14].

The obfuscation techniques offered by various developers have several gaps. The obfuscation techniques are able to protect the code to some extent; however, the code contains some debugging information. There is no obfuscator tool that can be completely declared as the best obfuscation technique. If the secret of an obfuscator is known, reverse engineers can easily accomplish their tasks by constructing de-obfuscators. These de-obscuscator tools have not yet been published, but in the future there may be the possibility of developing de-obscuscators [15].

The bytecode contains unknown characters and symbols from the source code. Reverse engineers have cracked the secrets of the byte code using reverse engineering tools. Therefore, it is possible to copy the original code after reversal, improve it, and resell it on the market to gain an advantage over the original author who developed the code in the first place. Some software development companies hire a hacker or a reverser to crack their code and find out the weaknesses and vulnerabilities of the software so that the company can fix it before it is actually hacked. All software programs contain a security key or registry file that ensures the protection of the software. Reverse engineers convert this file into source code when they remove the registration file from the software and use the exposed code for their own illegal development purpose [16].

Most of the obfuscation techniques are applied in the source file. These obfuscation techniques are applied individually in the source file. Most obfuscation techniques focus on renaming the identifiers and hiding the meaning of the code. Most reverse engineering tools are capable of analyzing the obfuscated code. According to the discussion in the papers highlighted in this research, they do not include mathematical equations to convert or encrypt the strings in the source file, and they do not include garbage conversion to change the layout of the code [17].

Obfuscation techniques are applied in the source code as a single technique. For example, the developer uses only variable names or hides only the names of classes. None of the papers discussed the use of a hybrid obfuscation technique, and none discussed a hybrid obfuscation technique with a mathematical equation for protection. For the obfuscation technique to be strong, it must be merged or joined. If the developer uses more than one obfuscation technique, there is a good chance that the code is protected from the reversal tools. The developer selects the obfuscation techniques that work together based on the layout and complexity of the original code. From the work examined in this study, the use of combined or hybrid obfuscation techniques guarantees strong protection against prohibited reverse engineering [18].

Table 1 presents limitations of most common obfuscation techniques.

List	Limitation
• Logistic map • Cipher block chaining • Symmetric cipher [19]	This technique uses mathematical equations to replace the text in the string with a chaos stream. The technique uses a secret key for encryption and uses a mathematical equation. The key can be randomly generated at the time of encryption or acquired from the developer. If the reverser can guess the key, there is the possibility of using the key to decrypt the entire code.
• Renaming • Hiding [20]	This technique emphasizes to hide features and change the layout. This is harder to understand but is not impossible to reverse. These tools can hide the code somehow. Nevertheless, reverse engineering is possible
Key hiding obfuscation [21]	This technique emphasizes to execute software and leave the source code and class file unchanged. Reversing tools have the ability to find and crack the key to the source file and perform code analysis.
Encryption [22]	This technique encrypts the executable code. The limitation of this technique is the programmer either limits key or round sizes, or leaves only stubs for restricted classes. Longer keys are used in encryption to provide better security. The longer key length in itself leads to slower encryption speed.
Packing [23]	This technique puts all the code into one package. The reversing tool is currently able to unpack the packed code and create new code that is useful and produces the same output as the original.
Classes combination obfuscation [24]	This technique hides classes by combining them. The inversion tools allow the user to create new classes and open the combined classes. The inversion tools contain great analysis function that allows the user to find the class trees and the connection between the classes.
Junk code obfuscation [25]	This technique emphasizes to change the names of the identifiers to create confusion while reading the code, and the reversing tools are able to create new names for the variables and classes by using characters. Then, the reverser can use the refactor function to create meaningful names.

Table 1.
Current obfuscation limitation.

3. Implementing hybrid obfuscation of encryption

In this section, we introduce a new hybrid obfuscation technique based on identifier renaming and string encryption. The technique relies on hybrid identifier renaming in the program's source file to cause extreme confusion for both reversal tools and humans when they examine the source file without permission. Regardless of the obfuscation strategy used, it was possible to contrast the obfuscation by renaming the identifiers and string encoding in two phases to first overcome the preemptive obfuscation and then add type information to the identifiers in the source code to contrast the layout obfuscation.

The first phase is renaming, and the hybrid obfuscation technique consists of two sections. The first section is obfuscating the identifiers to junk code to hide the meaning and increase complexity and confuse the decompiler during reversing. The second section is replacing the system keywords with Unicode.

The second phase is string encryption, where a set of random mathematical equations are injected into the strings to encrypt them. A transformation framework has been implemented to represent the steps of the hybrid obfuscation technique. The proposed technique can be used for many languages such as Arabic, English, Chinese, and so on. Using this technique creates the possibility of programming in different languages instead of English, which increases the protection of the code.

Following sections discuss the hybrid obfuscation encryption in detail:

3.1 Unicode approach

In the Java language, each character or symbol is represented using Unicode, which creates a possibility of changing the form of the code while reading. This technique is used in the source file. If this file is stolen, there will be no way to read it. The thief has to translate any Unicode to understand the meaning and figure out the code. The compiler is able to read Unicode and produce output. Combining Unicode with other encoding techniques in the source file makes it stronger. **Table 2** presents examples of Unicode [26].

0x0030	**0**	**0x0044**	**D**	**0x0051**	**Q**	**0x0064**	**d**	**0x0071**	**q**
0x0031	1	0x0045	E	0x0052	R	0x0065	e	0x0072	r
0x0032	2	0x0046	F	0x0053	S	0x0066	f	0x0073	s
0x0033	3	0x0047	G	0x0054	T	0x0067	g	0x0074	t
0x0034	4	0x0048	H	0x0055	U	0x0068	h	0x0075	u
0x0035	5	0x0049	I	0x0056	V	0x0069	i	0x0076	v
0x0036	6	0x004A	J	0x0057	W	0x006A	j	0x0077	w
0x0037	7	0x004B	K	0x0058	X	0x006B	k	0x0078	x
0x0038	8	0x004C	L	0x0059	Y	0x006C	l	0x0079	y
0x0039	9	0x004D	M	0x005A	Z	0x006D	m	0x007A	z
0x0041	A	0x004E	N	0x0061	a	0x006E	n	0x0A09	ਉ
0x0042	B	0x004F	O	0x0062	b	0x006F	o	0x0A0A	ਊ
0x0043	C	0x0050	P	0x0063	c	0x0070	p	0x2190	←
0x0A17	ਗ	0x2157	⅗	0x2175	vi	0x217F	m	0x2191	↑
0x0A18	ਘ	0x2158	⅘	0x2176	vii	0x2180	ↀ	0x2192	→
0x0A19	ਙ	0x2159	⅙	0x2177	vii	0x2181	ↁ	0x2193	↓
0x0AA	ਚ	0x215A	⅚	0x2178	ix	0x313A	ㄺ	0x33E1	㏡
0x0A1B	ਛ	0x215B	⅛	0x219E	↞	0x313B	ㄻ	0x33E2	㏢
0x1227	ሧ	0x215C	⅜	0x219F	↟	0x313C	ㄼ	0x33E3	㏣
0x1228	ረ	0x215D	⅝	0x21A0	↠	0x313D	ㄽ	0x33E4	㏤
0x1229	ሩ	0x215E	⅞	0x21A1	↡	0x313E	ㄾ	0x33E5	㏥
0x122A	ሪ	0x215F	⅟	0x21A2	↢	0x313F	ㄿ	0x33E6	㏦
0x122B	ራ	0x2160	I	0x21A3	↣	0x3140	ㅀ	0x33E7	㏧
0x122C	ሬ	0x2161	II	0x21A4	↤	0x3141	ㅁ	0x33E8	㏨
0x122D	ር	0x2162	III	0x21A5	↥	0x3142	ㅂ	0x33E9	㏩
0x122E	ሮ	0x2163	IV	0x21A6	↦	0x3143	ㅃ	0x33EA	㏪
0x2125	℥	0x2164	V	0x311D	ㄝ	0x33A2	㎢	0x33EB	㏫
0x2126	Ω	0x2165	VI	0x311E	ㄞ	0x33A3	㎣	0xA000	ꀀ
0x2127	℧	0x2166	VII	0x311F	ㄟ	0x33A4	㎤	0xA001	ꀁ

0x0030	**0**	**0x0044**	**D**	**0x0051**	**Q**	**0x0064**	**d**	**0x0071**	**q**
0x2128	ℨ	0x2167	Ⅶ	0x3120	ㄠ	0x33A5	㎥	0xA002	ꀂ
0x2129	℩	0x2168	Ⅸ	0x3121	ㄡ	0x33A6	㎦	0xA003	ꀃ
0x212A	K	0x2170	ⅰ	0x3122	ㄢ	0x33A7	㎧	0xA004	ꀄ
0x212B	Å	0x2171	ⅱ	0x3123	ㄣ	0x33A8	㎨	0xA005	ꀅ
0x2130	ℰ	0x2172	ⅲ	0x3124	ㄤ	0x33A9	㎩	0xA006	ꀆ
0x2131	ℱ	0x2173	ⅳ	0x3125	ㄥ	0x33AA	㎪	0xA007	ꀇ
0x2132	Ⅎ	0x2174	ⅴ	0x3126	ㄦ	0x33E0	㏠	0xA008	ꀈ

Table 2.
Uniocode characters.

In this approach, a Unicode transformation was used to rename the system keywords. The purpose of this renaming is to make the code in the source file more complicated. In this case, when reading the source file, the attacker will not be able to recognize the actual meaning of the code. This approach is very beneficial because in case of stealing the source file, the reader is not able to recognize the actual meaning of the code. He has to translate the whole code to understand the purpose of the code. However, even if the Unicode is easy to translate, the keywords of the system do not have much meaning, because the classes and variables in the functions and methods are.

3.2 String encryption approach

In this approach, a mathematical equation with a character field and loops were used to encode the strings in the source file. The encoding of the strings causes confusion while decompiling. The reversing tool is not able to translate the symbols generated by the mathematical equation; moreover, the compiler cannot translate the symbols that were converted to bytecode during compilation. The purpose of string encoding is to create a chaos stream in the source file and in the reverse file after decompiling [27]. The advantage of string encoding is that the mathematical formula used to create the chaos stream that can be used **N** times in the source code, and multiple (**X**) sets of mathematical equations can be used in the same source file. The more the chaos streams are created in the source file, the more the confusion is created during decompiling. The mathematical equations used in the source file were derived from the concept that Java programming language provides a function that can be used to convert the mathematical equation characters into different symbols. Normally, the equation contains a fixed value to ensure accurate output [28]. For the proposed technique, the value for the equation is two which will assigned to (**P**). There is other two values in the equation that are the values of (**Y**) and (**Z**). The values of (**Y**) and (**Z**) have to be carefully declared and assigned to produce the accurate output.

If the value of **Y** is **17** then the value of **Z** is **2**.

If the value of **Y** is **19** then the value of **Z** is **4**.

If the value of **Y** is **16** then the value of **Z** is **1**.

According to the above conditions, if the value of (**Y**) increases by one value, then the value of (**Z**) has to increase by one as well. The assigned value of (**P**) is 2, it can be changed as well to increment by one, and then the value of (**Y**) has to decrease by three values in order to get the calculation right for accurate output. The final result of calculating the three values have to be always 17; therefore, the value of (**P**) is fixed but it can decrease by one value, to increase the value of **(Y)** by one value as well. To prevent errors, the value of **(P)** was fixed at 2. The values of **(Y)** and **(Z)** can be increased and decreased accurately to allow using more mathematical equations in the source file. The final equation is:

$$Char = \frac{V}{2 + Y + Z}. \tag{3}$$

3.3 Mathematical equation to encrypt strings

The equation that was used to encrypt the strings in the source code is associated with beneficial attributes, and **(Y)** indicates the ideal (best) value of the considered attribute among the values of the attribute for different alternatives, and the fixed and best value for the equation is **2**; this value will not be changed. In the case of beneficial attributes for instance, those of which higher values are desirable for the given application, **(Y)** indicates the higher value of the attribute, and the highest value which will be used for the equation is **17** [29].

Lower values are desired for the given application, *and* **(Z)** indicates the lower value of the attribute. **(Z)** indicates the lowest value of the considered attribute among the values of the attribute for different alternatives, and the lowest value which will be used is **2**. In the case of beneficial attributes, **(Z)** indicates the lower value of the attribute. In the case of non-beneficial attributes, **(Y)** indicates the higher value of the attribute [30]. Following equation presents the string encryption transformation:

$$Char = \frac{V}{2 + Y + Z.} \tag{4}$$

3.4 Identifiers renaming to junk obfuscation

The main purpose of junk renaming is to create complicated code that is difficult to read and understand and make sense out of it. Junk renaming is used to confuse the reversing tool which leads to incorrect analysis and thus produces incorrect codes. Junk conversion provides the ability to create a variety of languages during the development of the software to protect it. The class file contains the junk code after compiling the source file. After using junk conversion, the converted code in the class file is converted back to junk code, which increases protection. Applying this feature means compromising some of the software quality factors that are readable code and manageable size. These features are compromised to increase the security of the code.

4. Hybrid obfuscation of encryption

Java development is based on object orientation, while the compiler executes the application based on components, unlike structured programs developed with the C programming language. Therefore, code obfuscation will not be a problem when compiling to machine language or bytecode. To use this hybrid obfuscation technique, certain steps must be followed. The first step is to use Object Junk Renaming Obfuscation [31–35]. This conversion must be done first to avoid confusion and errors when the obfuscation process is running. The second step is to encrypt strings. This technique must be performed second to have smooth conversion without errors. The last step is the Unicode renaming technique for obfuscation. Performing the hybrid obfuscation technique increases the security level of the code where reversing is nearly impossible. **Table 3** presents a sample of code after merging three approaches of obfuscation and after reversing.

There is possibility to change names to junk and can be used for any purposes such as emails, login, and so on. With this logic, the hybrid obfuscation encryption can be used to write encrypted letters and create a whole system using only junk code. **Table 4** presents names before and after obfuscation. Every time the obfuscated name is copied or used, it changes automatically.

The string encryption makes the obfuscation technique more effective in terms of securing the code, as it contains so many symbols that help to confuse the decompiler while parsing and analysis. **Figure 5** presents the framework of the proposed hybrid obfuscation encryption (**Figure 6**).

<table>
<tr><th>Obfuscated code</th><th>After reversing</th></tr>
<tr><td>\u0066\u006F\u0072\u0028 \u0069\u006E\u0074
\u046D01101: “ⴷÌÆÁÆ¾Ì¼˛Æ ˷Á¾Ë°¬¤Î¬ ´Ë ˛°¤¾¸
ÁÜ¤° É°⁻ ﺳÁÁ ⴷ Ü ÉÁ¼d d°²¤¸µ¼Õ |Æ ¸É¤Î¤ ?µ°°¤µ É°⁻
ﺳÁÁⴷ Ü ۴ ? מ ⴷ ﺳ בּ þ ? ?”.toCharArray\u0028\u0029
\u0029</td><td>BufferedWriter out = new BufferedWriter
(new FileWriter(de7, true));
Char c010101c[] = “\255\276 \313 \255 \306
\310\313\306\265\276\260 “.toCharArray();
int. 0908 = c010101c.length;
for(int d9 = 0; d9 < 0908; d9++)</td></tr>
<tr><td>\u0076\u006F\u0069\u0064龜\u00 28\u0029\u007B
\u0066\u006F\u00 72\u0028\u0069\u006E\u0074ьж01 1
靖\u003A”¤¼ ÀÌ¾Ë ë¬ ?ÀÈµË”\u00 2E\u0074\u006F
\u0043\u0068\u00 61\u0072\u0041\u0072\u0072\u00 61
\u0079\u0028\u0029\u0029\u00 7B\u0053\u0079\u0073
\u0074\u00 65\u006D\u002E\u006F\u0075\u00 74
\u002E\u0070\u0072\u0069\u006E\u0074\u0028\u0028
\u0063\u0068\u0061\u0072\u0029\u0028ьж011靖\u002F
\u0032\u002B\u0031\u0037\u002D\u0032\u0029\u0029
\u 003B\u007D \u0053\u0079\u0073\ u0074\u 0065
\u006D\u002E\u006F \u0075\u007 \u002E\u0070\u0072
\u0069\u006E\u0074\u0028”\n”\u 0029\u003B癩
\u003D更\u002E\u0 06E\u0065\u0078\u0074\u0044\u0
06F\u0075\u0062\u006C\u0065\u0028\u0029\u003B裸
\u003D裸\u002 B癩\u003B\u0066\u006F\u0072\u0 028
\u0069\u006E\u0074 ьж011精\u 003A”ÌÆÆ¬¾Ë⁼¤g¤¾
¬”\u002E\ u0074\u006F\u0043\u0068\u0061
\u0072Array\u0028\u0029\u0029\u007B\u0053\u0079
\u0073\u0074\u 0065\u006D\u002E\u006F\u0075\u
0074\u002E\u0070\u0072\u0069\u 006E\u0074\u0028
\u0028\u0063\u 0068\u0061\u0072\u0029\u0028ьж011精
\u002F\u0032\u002B\u0031\u0037\u002D\u0032\u0029
\u0029\ u003B\u007D\u0053\u0079\u0073\ u0074\u0065
\u006D\u002E\u006F\u0075\u0074\u002E\u0070\u0072
\u0069\u006E\u0074\u0028”\n”\u0029\u003B\u0053
\u0079\u0073\u0074\u0065\u006D\u002E\u006F\u0075
\u0074\u002E\u0070\u0072\u0069\u006E\u0074\u0028
裸u0029\u003B\u007D\u0076 \u006F\u0069 \u0064 契
\u0028\u0029\u007 B</td><td>After reversing the class file which contains
full code

package bankencrypt;
// Referenced classes of package
bankencrypt:
// F9A4
public class Bankencrypt
{
public Bankencrypt()
{
}
public static void main (String args[])
{
F9A4 A461 = new F9A4();
A461.F907();
A461.F908();
A461.F909();
A461.F90A();
}
}</td></tr>
</table>

Table 3.
Obfuscated code before and after reversing.

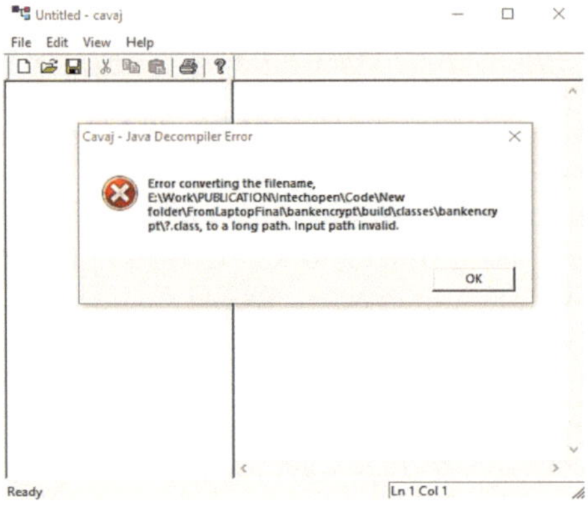

Figure 5.
Reversing hybrid obfuscation.

Name	After obfuscation
Asma mahfoud	ÉÁ¼d d°²¤¸µ¼Õ
Java hacker	\"v¤Î¤\"²¤¨¸¬Æ
Kesava	¸É¤Î¤
hi	³׳
keep it real	µË Æ¤°
Nur	⅄Ì Æ

Table 4.
Names before and after obfuscating.

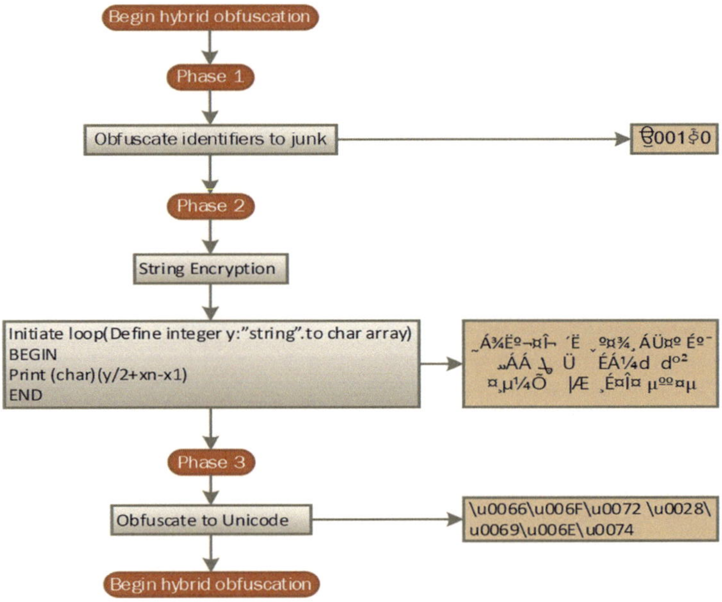

Figure 6.
Hybrid obfuscation encryption framework.

5. Empirical evaluation of the hybrid obfuscation

Four reversing tools were used to test the effectiveness of the technique and to determine how much can the reversing tool uncover and read from the obfuscated code. Four reversing tools were used for this experiment; the tools are CAVAJ, JAD, DJ, and JD. The parameters are distributed among the reversing tools based on their behavior toward the obfuscated code. For instance, JD only tested the identifiers names because it has the ability to reveal the entire code; therefore, there was no need to test the rest of parameters. **Figure 7** presents experiment design.

5.1 Testing with CAVAJ

CAVAJ as reversing tool for Java class file *is* used to determine the ability of it to read the code after obfuscating. **Figure 8** presents the results of CAVAJ testing.

5.2 Testing with Java decompiler (JD)

DJ Reversing tool is used to determine the ability to reverse Java class file that contains hybrid obfuscated technique. The test will determine if the tool is able to read the obfuscated code, and how much can the tool read and discover. **Figure 9** presents the output after reversing.

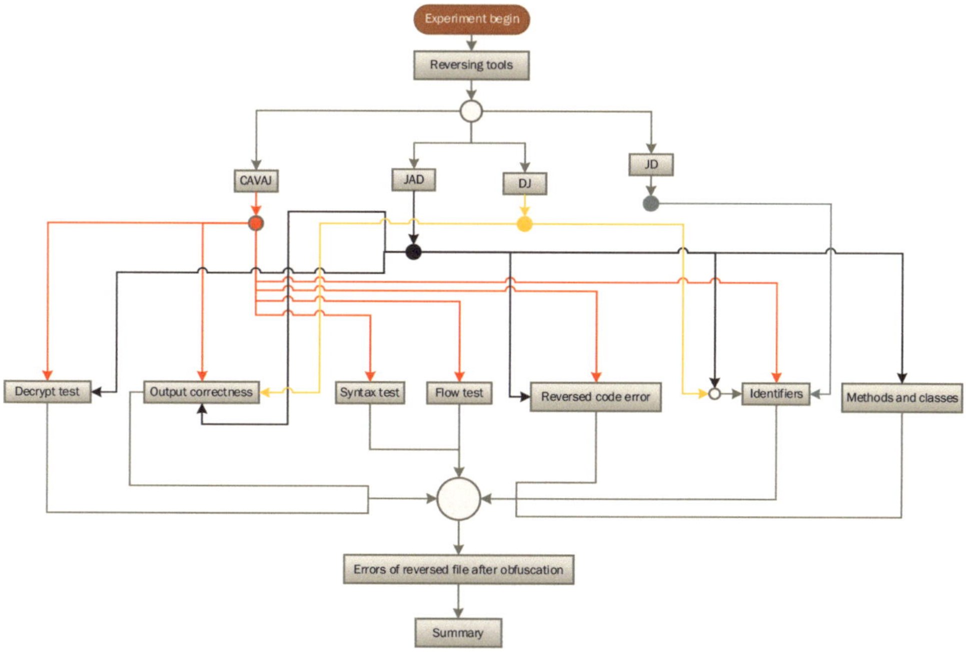

Figure 7.
Experiment design.

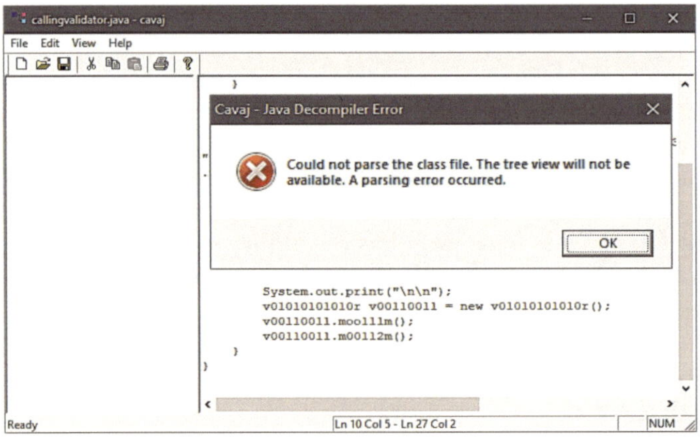

Figure 8.
Reversing result of CAVAJ.

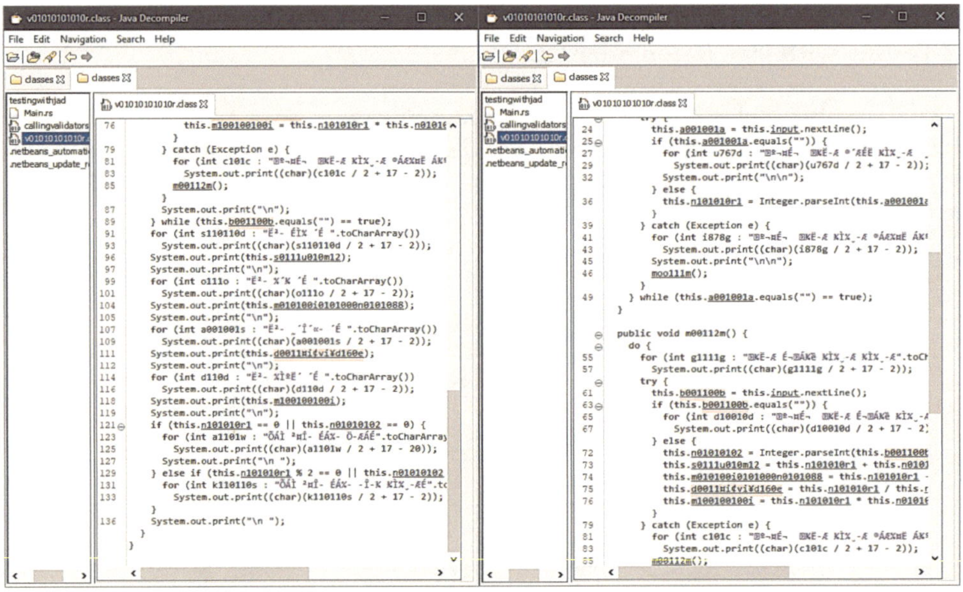

Figure 9.
Reversing result of JD.

5.3 Testing with JAD

After installing JAD, prompt command is used to find the Java class file, then the file is opened in command, and the file name.jad is typed to reverse the file. **Figure 10** presents the result of reversing.

First and second classes test for output correctness and reversed code error:

The tool was not able the code after obfuscation with hybrid technique, and it has presented errors while reading and just revealed the Unicode without the ability to read the identifiers.

First and second classes test for methods and classes and identifiers:

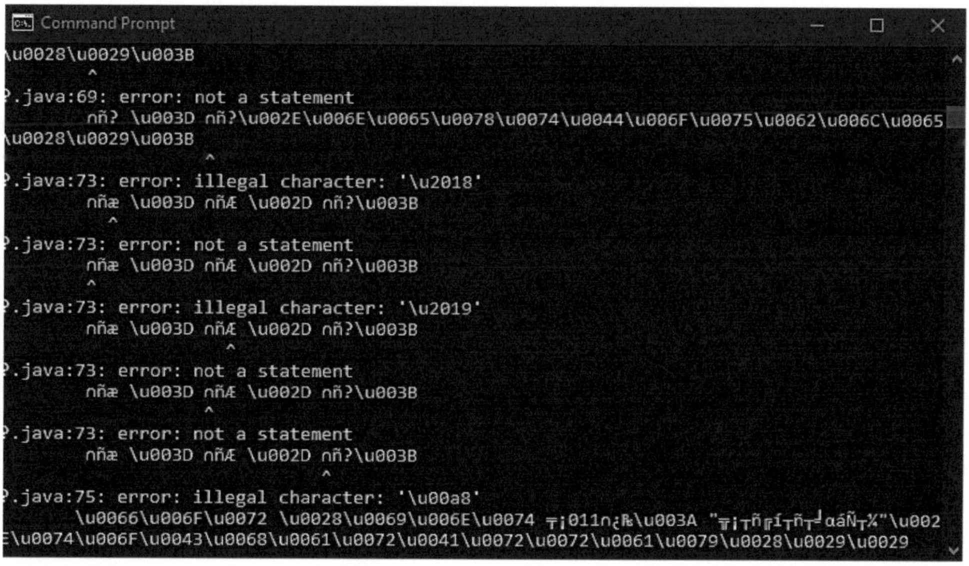

Figure 10.
Reversing result with JAD.

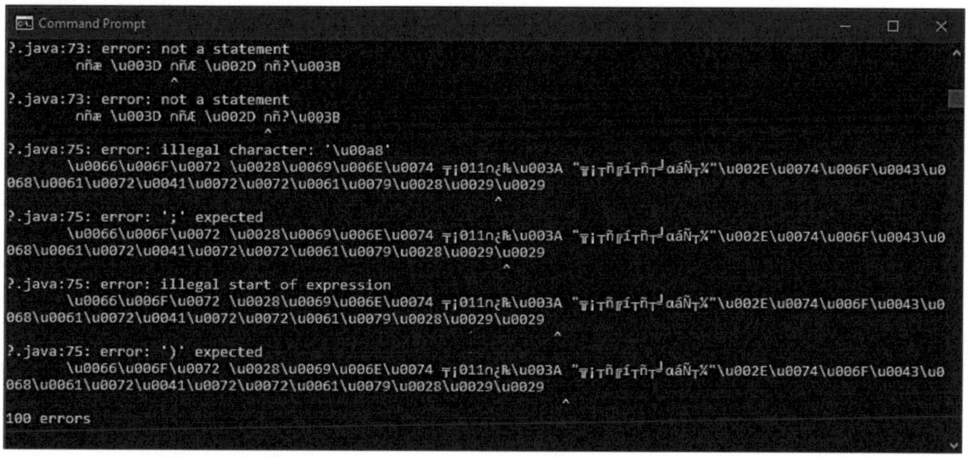

Figure 11.
JAD reversing result for methods and identifiers.

Based on **Figure 11**, the tool was not able to get a meaning of the encrypted strings and identifiers; in fact, it has changed the names further which can be considered for the another level of protection. This way the reverser will not be able to read the code or get a meaning of it, and also the name of the Java file was encrypted to mislead the reverser if the source file is stolen. **Figure 12** presents the form of the Java file name after encryption.

5.4 Testing to Decompiler java (DJ)

DJ reversing tool Java is a tool that reverses the class file. This tool is used to determine the ability to reverse Java class file that contains hybrid obfuscated

Figure 12.
File name after encryption.

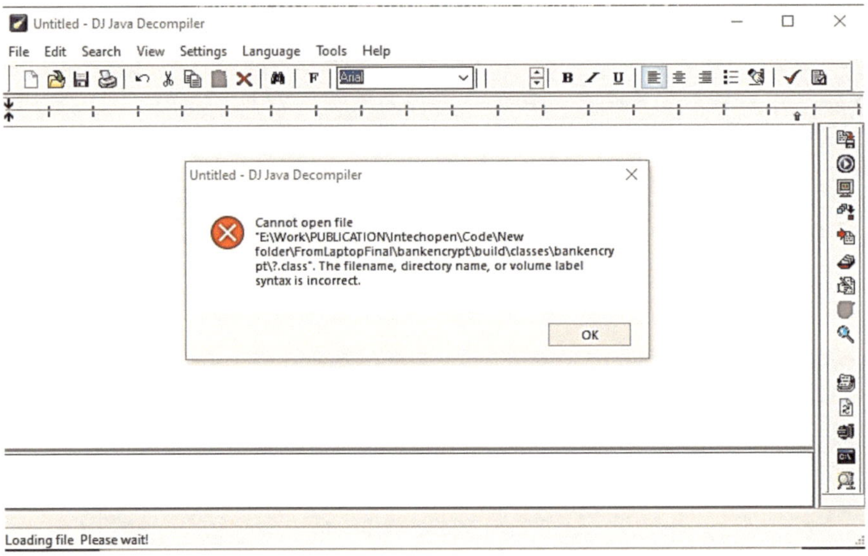

Figure 13.
Reversing result with DJ.

technique. The test will determine if the tool is able to read the obfuscated code, and how much can the tool reveal. **Figure 13** presents the reversing result of reversing the class file of output correctness.

First class test/output correctness.

The tool was not able to read the first-class test to reveal the code. Therefore, there is no code to test its correctness. This is promising results of having hybrid obfuscation technique. An error message is appeared to define syntax error.

Second class test/identifiers.

According to **Figure 14**, the tool was not able to read the code after obfuscation. This results the proof that using hybrid obfuscation is more beneficial than just applying one technique.

6. Conclusion

The hybrid obfuscation technique was effective to protect the code. The reversing tools were not able to read and translate the encrypted strings. Renaming to junk in the obfuscation technique was effective as the reversing tool has converted the junk to a series of random numbers and symbols. The reversing tool was able to read the

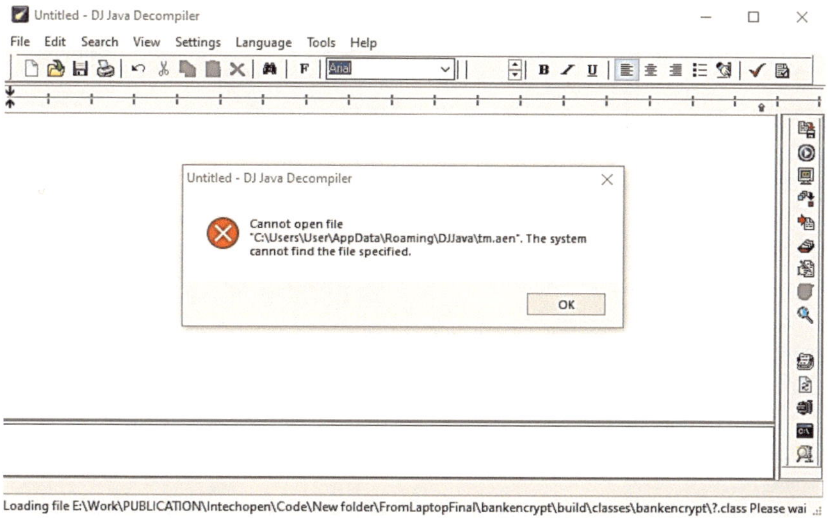

Figure 14.
Identifiers test.

system keywords only. Furthermore, the reversing tool has added methods and preprocessors while parsing the file. The reversing tool was not able to analyze the obfuscated code to get appropriate output. This means that the hybrid obfuscation technique is effective to protect the source file from prohibited reverse engineering. Third objective of this research was successfully met; according to the experimentation, a series of junk and chaos was created after reversing the obfuscated code.

The extreme chaos was generated due to the merge of string encryption and renaming approaches in one source file which has led to confusion while reversing as the reversing tool was not able to translate or read or analyze the code. To summarize the results of the experiments that were conducted before and after obfuscation, we calculate the lines of code (LOC) of original file before and after reversing, calculate the total errors appeared during running the reversed file before and after obfuscation, and then find the difference to determine the strength.

Based on the results of the reversing tools, they were not able to discover fully functioning code; in all cases, the reversing tools have generated a series of chaos and random numbers and symbols while attempting to translate the obfuscated code. The code that was generated from the reversing tools did not provide an output, and there was always an error while trying to compile the obfuscated code after reversing.

Table 5. The summary of errors occurred for the four tested cases.

Reversing tool	Testing component	Reversed file before hybrid technique	Reversed file after hybrid technique
CAVAJ	Compiled reversed code error test	Zero	6
	De-Crypt String test		1
JAD	Output correctness		7

Reversing tool	Testing component	Reversed file before hybrid technique	Reversed file after hybrid technique
	Compiled reversed code error test		100
	Methods and classes correctness test		22
DJ	Output correctness test		0
JD	Identifiers names test		0

Table 5.
Error summary.

7. Future work

The number and type of obfuscators we used for our research were fairly small. Future work could explore a wider variety of noncommercial and research obfuscators to provide a broader picture of protection possibilities. Due to time constraints, we were also not able to take advantage of all commercial obfuscators that we had access to. In the future, more commercial obfuscators and reversing tools can be used for the sake of this research. The proposed hybrid obfuscation technique can be further used for games and mobile applications to protect financially from being illegally reversed.

The technique can be developed with C/C++ programming language instead of Java, as Java is closer to the hardware level and communicate with it easily due to the pointer feature it has. Having the technique implemented with C/C++ is an advantage which makes the tool stronger for more defensive.

The technique can be as an added tool in the programming environment such as NetBeans or eclipse where programmer can customize which part of the code to be encrypted and which approach to use. Programmer has full freedom to mix and match encryption approaches in the code to increase security. Having such encryption tool prevents errors while encryption and saves time.

The proposed technique's concept can be used in any programming language to what fits its requirements and mechanisms and also opens an opportunity to have an option to insert different verbal languages, such as Arabic, Chinese, or any other language, for the sake of encryption to increase the level of security.

Author details

Asma'a Al-Hakimi[1*] and Abu Bakar Md Sultan[2]

1 Faculty of Information Sciences and Engineering, Management and Science University, University Drive, Off Persiaran Olahraga, Shah Alam, Selangor Darul Ehsan, Malaysia

2 Department of Software Engineering and Information System, Faculty of Computer Science and Information Technology, Universiti Putra Malaysia (UPM), Serdang, Malaysia

*Address all correspondence to: selfemoon@gmail.com

References

[1] Yasin A, Nasra I, Yasin A, Nasra I. Dynamic multi levels Java code obfuscation technique (DMLJCOT). International Journal of Computer Science and Security (IJCSS). 2016;**10** (4):140-160

[2] Kumar R, Vaishakh ARE. Detection of obfuscation in Java malware. Physics Procedia. 2016;**78**(2015):521-529

[3] Darwish SM, Guirguis SK, Zalat MS. Stealthy code obfuscation technique for software security. In: Proceedings, the 2010 International conference on computer engineering & systems (ICCES'2010). Cairo, Egypt: IEEE; 2010. pp. 93-99

[4] Cho T, Kim H, Yi JH. Security assessment of code obfuscation based on dynamic monitoring in android things. IEEE Access. 2017;**5**:6361-6371

[5] Xiang G, Cai Z. The code obfuscation technology based on class combination. In: Proc. DCABES 2010 ninth international symposium on distributed computing and applications to business, engineering and science. Hong Kong, China: IEEE. Vol. 60970064, 2010. pp. 479-483

[6] Deshmukh GC, Patil SM. Study for best data obfuscation techniques using multi-criteria decision-making technique. International Journal of Computer Applications. 2018;**180**(43):50-57

[7] Al-Hakimi AMH, Sultan ABM, Ghani AAA, Ali NM, Admodisastro NI. Hybrid obfuscation technique to protect source code from prohibited software reverse engineering. IEEE Access. 2020; **8**:187326-187342

[8] Sebastian SA, Malgaonkar S, Shah P, Kapoor M, Parekhji T. A study & review on code obfuscation. IEEE WCTFTR 2016. In: Proc. 2016 World Conf. Futur. Trends Res. Innov. Soc. Welf. Coimbatore, India: IEEE; 2016

[9] Batchelder M, Hendren L. Obfuscating Java: The most pain for the least gain. Lecture Notes in Computer Science. 2007;**4420**:96-110

[10] Peng Y, Chen Y, Shen B. An adaptive approach to recommending obfuscation rules for Java bytecode obfuscators. In: Proceedings, 2019 IEEE 43rd annual computer software and applications conference (COMPSAC). Vol. 1. Milwaukee, WI, USA: IEEE; 2019. pp. 97-106

[11] Kumar C, Bhaskari DL. Different obfuscation techniques for code protection. 4th International Conference on Eco-friendly Computing and Communication Systems. 2015;**70**: 757-763

[12] Ceccato M et al. Towards experimental evaluation of code obfuscation techniques. In: Proceedings, CCS08: 15th ACM conference on computer and communications security, Alexandria, Virginia, USA: ACM; 2008. pp. 39-45

[13] Solomonoff RJ. Algorithmic probability: Theory and applications. Information Theory and Statistical Learning. New York: Springer; 14 Nov 2008:1-23. ISBN 978-0-387-84815-0; 978-1-4419-4650-8; 978-0-387-84816-7. DOI: 10.1007/978-0-387-84816-7

[14] Budhkar S. Reverse engineering Java code to class diagram: An experience report. International Journal of Computer Applications. 2011;**29**(6): 36-43

[15] Baxter ID, Mehlich M. Reverse engineering is reverse forward engineering. Science of Computer Programming. 2000;**36**(2):131-147

[16] J. M. Memon, Shams-ul-Arfeen, A. Mughal, and F. Memon, Preventing reverse engineering threat in java using byte code obfuscation techniques. In: Proceedings, International conference on emerging technologies, ICET 2006, November. Peshawar, Pakistan: IEEE; 2006. pp. 689–694

[17] Zhang L, Meng H, Thing VLL. Progressive control flow obfuscation for android applications. In: TENCON 2018 - 2018 IEEE Region 10 Conference. Vol. 2018. Jeju, Korea (South): IEEE; 2019. pp. 1075-1079

[18] You I. Malware Obfuscation Techniques: A Brief Survey. Fukuoka, Japan: IEEE; 2010. pp. 297-300

[19] Popa M. Techniques of program code obfuscation for secure software. Journal of Mobile, Embedded and Distributed Systems. 2011;**III**(4):205-219

[20] Tang Z, Chen X, Fang D, Chen F. Research on java software protection with the obfuscation in identifier renaming. In: 2009 Fourth International Conference on Innovative Computing, Information and Control (ICICIC), Kaohsiung, Taiwan: IEEE; 2009. Vol. 2009. 2007. pp. 1067-1071. DOI: 10.1109/ICICIC.2009.312. ISBN: 978-1-4244-5544-7; 978-1-4244-5543-0; 978-0-7695-3873-0

[21] L. Luo, J. Ming, D. Wu, P. Liu, and S. Zhu, Semantics-based obfuscation-resilient binary code similarity comparison with applications to software plagiarism detection. In: Proc. 22nd ACM SIGSOFT Int. Symp. Found. Softw. Eng. - FSE 2014, IEEE. 2014;**43** (12):389–400. Available from: https://ieeexplore-ieee-org.ezadmin.upm.edu.my/document/7823022

[22] Bergström E, Åhlfeldt RM. Foundations and practice of security. In: Lect. Notes Comput. Sci. (Including Subser. Lect. Notes Artif. Intell. Lect. Notes Bioinformatics). Switzerland: Springer Verlag. Vol. 9482. 2016. pp. 268-276

[23] Amalarethinam DIG, Geetha JS. Image encryption and decryption in public key cryptography based on MR. In: Proc. Int. Conf. Comput. Commun. Technol. ICCCT 2015. Chennai, India: IEEE. June. 2015. pp. 133-138

[24] Iman M u, Ishaq AFM. Anti-reversing as a tool to protect intellectual property. In: Eng. Syst. Manag. Its Appl. (ICESMA), 2010 Second Int. Conf. Sharjah, United Arab Emirates: IEEE; 2010. pp. 1-5

[25] Angyal L, Lengyel L, Charaf H. An overview of the state-of-the-art reverse engineering techniques. In: 7th International Symposium of Hungarian Researchers on Computational Intelligence. Budapest, Hungary: HUCI; 2006. pp. 507-516

[26] Leahy P. What is unicode?, ThoughtCo. 2017. p. 1

[27] Real-time MAT, Ef- PCCS, Butaha MA. Crypto-compression systems for efficient embedded to cite this version. [thèse de Doctorat']. 2017

[28] Wang ZY, Wu WM. Technique of javascript code obfuscation based on control flow tansformations. Applied Mechanics and Materials. 2014;**519–520** (Iccse):389-392

[29] Sun Y. How to render mathematical symbols in Java. March, 2003

[30] Baker SIB, Al-Hamami AH. Novel algorithm in symmetric encryption (NASE): Based on feistel cipher. In: Proc. 2017 International Conference on New Trends in Computing Sciences (ICTCS), 2017. Amman, Jordan: IEEE; Jan 2018. Vol. 3. 2017. pp. 191-196

[31] Sosonkin M, Naumovich G, Memon N. Obfuscation of design intent in object-oriented applications. In: DRM 2003 Proc. 3rd ACM workshop on Digital rights management, Washington DC USA: Association for Computing Machinery. 2003. pp. 142-153

[32] Alkawaz MH, Steven SJ, Hajamydeen AI. Detecting phishing website using machine learning. In: 2020 16th IEEE International Colloquium on Signal Processing & its Applications (CSPA). Langkawi, Malaysia. 2020. pp. 111-114. DOI: 10.1109/CSPA48992.2020.9068728

[33] Al Yahyaee OMAR. Information Security Management in Abu Dhabi Police, UAE. [Doctoral dissertation] Management & Science University. 2016

[34] Alkawaz MH, Steven SJ, Hajamydeen AI, Ramli R. A Comprehensive survey on identification and analysis of phishing website based on machine learning methods. In: 2021 IEEE 11th IEEE Symposium on Computer Applications & Industrial Electronics (ISCAIE). Penang, Malaysia. 2021. pp. 82-87. DOI: 10.1109/ISCAIE51753.2021.9431794

[35] Alkawaz MH, Joanne Steven S, Mohammad OF, Gapar Md Johar M. Identification and analysis of phishing website based on machine learning methods. In: 2022 IEEE 12th Symposium on Computer Applications & Industrial Electronics (ISCAIE). Penang, Malaysia. 2022. pp. 246-251. DOI: 10.1109/ISCAIE54458.2022.9794467

Chapter 5

Extended Intuitionistic Fuzzy Line Graphs: Theory and Properties

Venkata Naga Srinivasa Rao Repalle, Keneni Abera Tola and Maamo Abebe Ashebo

Abstract

The introduction of fuzzy set theory was given by Zadeh. The introduction of fuzzy graph theory was given by Kauffman. Later the structure of fuzzy graph was developed Rosenfeld. The traditional fuzzy set cannot be used to completely describe all the evidence in problems where someone wants to know in how much degree of non-membership. Such a problem got the solution by Atanassov who introduced intuitionistic fuzzy set which described by a membership, a non-membership and a hesitation functions. An intuitionistic fuzzy set is used to solve problems involving uncertainty and imprecision that can't be handled by a traditional fuzzy set. This chapter introduced the interval-valued intuitionistic fuzzy line graphs (IVIFLG) and explored the results related to IVIFLG. As a result, many theorems and propositions related to IVIFLG are developed and supported by proof. Moreover, some remarkable isomorphic properties, strong IVIFLG, and complete IVIFLG have been investigated, and the proposed concepts are illustrated with the examples.

Keywords: fuzzy set, interval-valued intuitionistic fuzy graph, interval-valued intuitionistic fuzzy line graph, isomorphism, isomorphic properties

1. Introduction

Since Euler was presented with the impression of the Königsberg bridge problem, graph theory has received recognition in a variety of academic fields, including natural science, social science, engineering, and medical science. In the field of graph theory, some operations such as the Wiener index of graphs, line graphs, total graphs, cluster and corona operations of graphs, edge join of graphs, and semi-total line have been useful. In addition, some properties of boiling point, heat of evaporation, surface tension, vapor pressure, total electron energy of polymers, partition coefficients, ultrasonic sound velocity, and internal energy can be analyzed in chemical graph theory. These operations are not only useful in classical graphs but also in fuzzy graphs and generalizations of fuzzy graphs. Because real-world problems are frequently fraught with uncertainty and imprecision, Zadeh proposed fuzzy sets and membership degrees [1]. Accordingly, Kaufman presented the concept of fuzzy relations based on Zedeh's work in [2]. Rosenfeld [3] assembled both Zedeh's and Kaufman's work and then introduced fuzzy graphs.

Later on, Atanassov observed that fuzzy sets (FS) did not handle many problems with uncertainty and imprecision [4]. Based on these observations, he combined the membership degree with the falsehood degree and presented intuitionistic fuzzy sets (IFS) with relations and IFG, which is a generalization of FS [4–6]. It has many applications in fuzzy control, and defuzzification is the most computationally intensive part of fuzzy control. Mordeson investigated the concept of fuzzy line graphs (FLG) for the first time and explored both sufficient and necessary conditions for FLG to be a bijective homomorphism to its FG. He developed some theorems and propositions [7]. Firouzian et.al [8] introduced the notion of degree of an edge in fuzzy line graphs and congraphs.

Akram and Dudek discussed interval valued fuzzy graph (IVFG) and its properties in [9]. Later, different classes of IVIFGs such as regular, irregular, highly irregular, strongly irregular and neighbourly irregular IVIFGs were discussed [10]. Then, Akram drived IVFLG from IVFG [11]. Interval-valued intuitionistic $(S,T)-$fuzzy graphs were introduced by Rashmanlou and Borzooei [12]. Afterward, the idea of intuitionistic fuzzy line graph (IFLG) studied by Akram and Davvaz [13]. Furthermore, IFLG and its properties are investigated in [14].

Based on the defined concepts, we gave the definition of IVIFLG in this chapter. Our works are novel in the following ways: (1) IVIFLG is presented and illustrated with an example, (2) numerous theorems and propositions are developed and proved; (3) further, interval-valued intuitionistic weak line isomorphism and interval-valued intuitionistic weak vertex homomorphism are proposed. Readers should refer [5, 7, 11] for notations that are not declared in this chapter.

2. Discussion

This section contains some basic definitions used to introduce IVIFLG. Throughout this chapter we considered only simple graph.

Definition 1.1. The graph $G = (V, E)$ is an intuitionistic fuzzy graph (IFG) if the following conditions are satisfied [15]

i. $\sigma_1 : V \to [0,1]$ and $\gamma_1 : V \to [0,1]$ are membership and nonmembership value of vertex set of G respectively and $0 \leq \sigma_1(v) + \gamma_1(v) \leq 1 \ \forall v \in V$,

ii. $\sigma_2 : V \times V \to [0,1]$ and $\gamma_2 : V \times V \to [0,1]$ are membership and nonmembership with $\sigma_2(v_iv_j) \leq \sigma_1(v_i) \wedge \sigma_1(v_j)$ and $\gamma_2(v_iv_j) \leq \gamma_1(v_j) \vee \gamma_1(v_j)$ and $0 \leq \sigma_1 2(v_iv_j) + \gamma_2(v_iv_j) \leq 1, \forall v_iv_j \in E$.

Definition 1.2. The line graph L(G) of graph G is defined as any node in $L(G)$ that corresponds to an edge in G, and pair of nodes in $L(G)$ are adjacent if and only if their correspondence edges $e_i, e_j \in G$ share a common node $v \in G$.

Definition 1.3. For the given graph $G = (V, E)$ with $n-$vertices and $S_i = \left\{v_i, e_{i_1}, \cdots, e_{i_p}\right\}$ such that $1 \leq i \leq n, 1 \leq j \leq p_i$ and $e_{ij} \in E$ has v_i as a vertex. Then (S, T) is called intersection graph where $S = \{S_i\}$ is the vertex set of (S, T) and $T = \{S_iS_j | S_i, S_j \in S; S_i \cap S_j \neq \varnothing, \text{for } i \neq j\}$ is an edge set of (S, T).

Definition 1.4. The line(edge) graph $L(G) = (H, J)$ is where $H = \{\ \{e\} \cup \{u_e, v_e\} : e \in E, u_e, v_e \in V, e = u_ev_e$ and $J = \{S_eS_f :\ e,f \in E, e \neq f, S_e \cap S_f \neq \varnothing\}$ with $S_e = \{e\} \cup \{u_e, v_e, e \in E\}$ [11].

Definition 1.5. Let $G = (A_1, B_1)$ is an IFG with $A_1 = (\sigma_{A_1}, \gamma_{A_1})$ and $B_1 = (\sigma_{B_1}, \gamma_{B_1})$ be IFS on V and E respectively. Then $(S, T) = (A_2, B_2)$ is an intuitionistic fuzzy intersection graph of G whose membership and nonmembership functions are defined as [14]

i. $\sigma_{A_2}(S_i) = \sigma_{A_1}(v_i), \qquad \gamma_{A_2}(S_i) = \gamma_{A_1}(v_i), \forall S_i, S_j \in S$

ii. $\sigma_{B_2}(S_i\, S_j) = \sigma_{B_1}(v_i v_j), \qquad \gamma_{B_2}(S_i S_j) = \gamma_{B_1}(v_i v_j) \forall S_i S_j \in T.$

where $A_2 = (\sigma_{A_2}, \gamma_{A_2})$, $B_2 = (\sigma_{B_2}, \gamma_{B_2})$ on S and T respectively. So, IFG of the intersection graph (S, T) is isomorphic to G(means, $(S, T) \cong G$).

Definition 1.6. Consider $L(G^*) = (H, J)$ be line graph of $G^* = (V, E)$. Let $G = (A_1, B_1)$ be IFG of G^* with $A_1 = (\sigma_{A_1}, \gamma_{A_1})$ and $B_1 = (\sigma_{B_1}, \gamma_{B_1})$ be IFS on X and E receptively. Then we define the intuitionistic fuzzy line graph $L(G) = (A_2, B_2)$ of G as

i. $\sigma_{A_2}(S_e) = \sigma_{B_1}(e) = \sigma_{B_1}(u_e v_e),$

$\gamma_{A_2}(S_e) = \gamma_{B_1}(e) = \gamma_{B_1}(u_e v_e)$, for all $S_e, S_e \in H$

ii. $\sigma_{B_2}(S_e S_f) = \sigma_{B_1}(e) \wedge \sigma_{B_1}(f)$

$\gamma_{B_2}(S_e S_f) = \gamma_{B_1}(e) \vee \gamma_{B_1}(f), \forall S_e S_f \in J..$

where $A_2 = (\sigma_{A_2}, \gamma_{A_2})$ and $B_2 = (\sigma_{B_2}, \gamma_{B_2})$ are IFS on H and J respectively.

The $L(G) = (A_2, B_2)$ of IFG G is always IFG.

Definition 1.7. Let $G_1 = (A_1, B_1)$ and $G_2 = (A_2, B_2)$ be two IFGs. The homomorphism of $\psi : G_1 \to G_2$ is mapping $\psi : V_1 \to V_2$ such that [14].

i. $\sigma_{A_1}(v_i) \le \sigma_{A_2}(\psi(v_i)), \quad \gamma_{A_1}(v_i) \le \gamma_{A_2}(\psi(v_i))$

ii. $\sigma_{B_1}(v_i, v_j) \le \sigma_{B_2}(\psi(v_i)\psi(v_j)),$

$\gamma_{B_1}(v_i, v_j) \le \gamma_{B_2}(\psi(v_i)\psi(v_j)) \ \forall v_i \in V_1, v_i v_j \in E_1.$

Definition 1.8. The interval valued FS A is characterized by [9].

$$A = \{v_i, [\sigma_A^-(v_i), \sigma_A^+(v_i)] : v_i \in X\}.$$

Here, $\sigma_A^-(v_i)$ and $\sigma_A^+(v_i)$ are lower and upper interval of fuzzy subsets A of X respectively, such that $\sigma_A^-(v_i) \le \sigma_A^+(v_i) \ \forall v_i \in V$.

For simplicity, we used IVFS for interval valued fuzzy set.

Definition 1.9. Let $A = \{[\sigma_A^-(v), \sigma_A^+(v)] : v \in X\}$ be IVFS. Then, the graph $G^* = (V, E)$ is called IVFG if the following conditions are satisfied;

$$\sigma_B^-(v_i v_j) \le (\sigma_A^-(v_i) \wedge \sigma_A^-(v_j))$$
$$\sigma_B^+(v_i v_j) \le \sigma_A^+(v_i) \wedge \sigma_A^+(v_j)$$

$\forall v_i, v_j \in V, \ \forall v_i v_j \in E$ and where $A = [\sigma_A^-, \sigma_A^+]$, $B = [\sigma_B^-, \sigma_B^+]$ is IVFS on V and E respectively.

Definition 1.10. Let $G = (A_1, B_1)$ be simple IVFG. Then we define IVF intersection graph $(S, T) = (A_2, B_2)$ as follows:

1. A_2 and B_2 are IFS of S and T respectively,

2. $\sigma^-_{A_2}(S_i) = \sigma^-_{A_1}(v_i)$ and $\sigma^+_{A_2}(S_i) = \sigma^+_{A_1}(v_i), \forall S_i, S_j \in S$ and

3. $\sigma^-_{B_2}(S_iS_j) = \sigma^-_{B_1}(v_iv_j), \sigma^+_{B_2}(S_iS_j) = \sigma^+_{B_1}(v_iv_j), \quad \forall S_iS_j \in T.$

Remark: The given IVFG G and its intersection graph (S, T) are always isomorphic to each other.

Definition 1.11. An interval valued fuzzy line graph (IVFLG) $L(G) = (A_2, B_2)$ of IVFG $G = (A_1, B_1)$ is defined as follows [11]:

- A_2 and B_2 are IVFS of H and J respectively, where $L(G^*) = (H, J)$
- $\sigma^-_{A_2}(S_i) = \sigma^-_{B_1}(e) = \sigma^-_{B_1}(u_ev_e), \ \sigma^+_{A_2}(S_i) = \sigma^+_{B_1}(e) = \sigma^+_{B_1}(u_ev_e),$
- $\sigma^-_{B_2}(S_eS_f) = \sigma^-_{B_1}(e) \wedge \sigma^-_{B_1}(f), \ \sigma^+_{B_2}(S_eS_f) = \sigma^+_{B_1}(e) \wedge \sigma^+_{B_1}(f)$ for all $S_e, S_f \in H, S_eS_f \in J.$

Definition 1.12. A graph $G = (A, B)$ with underlying fuzzy set V is IVIFG if

i. the mapping $\sigma_A : V \to [0,1]$ and $\gamma_A : V \to [0,1]$ where $\sigma_A(v_i) = [\sigma^-_A(v_i), \sigma^+_A(v_i)]$ and $\gamma_A(v_i) = [\gamma^-_A(v_i), \gamma^+_A(v_i)]$ denote a membership degree and non membership degree of vertex $v_i \in V$, receptively such that $\sigma^-_A(v_i) \leq \sigma^+_A(v_i)$, $\gamma^-_A(v_i) \leq \gamma^+_A(v_i)$ and $0 \leq \sigma^+_A(v_i) + \gamma^+_A(v_i) \leq 1 \ \forall v_i \in V$,

ii. the mapping $\sigma_B : V \times V \subseteq E \to [0,1]$ and $\gamma_B : V \times V \subseteq E \to [0,1]$ where $\sigma_B(v_iv_j) = [\sigma^-_B(v_iv_j), \sigma^+_B(v_iv_j)]$ and $\gamma_B(v_iv_j) = [\gamma^-_B(v_iv_j), \gamma^+_B(v_iv_j)]$ such that

$$\sigma^-_B(v_iv_j) \leq \sigma^-_A(v_i) \wedge \sigma^-_A(v_j, \qquad \sigma^+_B(v_iv_j) \leq \sigma^+_A(v_i) \wedge \sigma^+_A(v_j)$$
$$\gamma^-_B(v_iv_j) \leq \gamma^-_A(v_i) \vee \gamma^-_A(v_j), \qquad \gamma^+_B(v_iv_j) \leq \gamma^+_A(v_i) \vee \gamma^+_A(v_j)$$

where $0 \leq \sigma^+_B(v_iv_j) + \gamma_B + (v_iv_j) \leq 1$ and $\forall v_iv_j \in E$.

In the next section, we begin the main findings of this chapter by introducing and demonstrating examples of IVIFLG.

Definition 1.13. Consider $L(G) = (H, J)$ is IVIFLG of IVIFG $G = (A_1, B_1)$ and denoted by $L(G) = (A_2, B_2)$ whose membership and non membership function is defined as

i. A_2 and B_2 are IVIFS of H and J respectively, such that

$$\sigma^-_{A_2}(S_e) = \sigma^-_{B_1}(e) = \sigma^-_{B_1}(u_ev_e)$$
$$\sigma^+_{A_2}(S_e) = \sigma^+_{B_1}(e) = \sigma^+_{B_1}(u_ev_e)$$
$$\gamma^-_{A_2}(S_e) = \gamma^-_{B_1}(e) = \gamma^-_{B_1}(u_ev_e)$$
$$\gamma^+_{A_2}(S_e) = \gamma^+_{B_1}(e) = \gamma^+_{B_1}(u_ev_e) \ \ \forall S_e \in H.$$

ii. The edge set of L(G) is

$$\sigma^-_{B_2}(S_eS_f) = \sigma^-_{B_1}(e)\wedge\sigma^-_{B_1}(f), \qquad \sigma^+_{B_2}(S_eS_f) = \sigma^+_{B_1}(e)\wedge\sigma^+_{B_1}(f)$$

$$\gamma^-_{B_2}(S_eS_f) = \sigma^-_{B_1}(e)\vee\gamma^-_{B_1}(f), \qquad \gamma^+_{B_2}(S_eS_f) = \gamma^+_{B_2}(e)\vee\gamma^+_{B_1}(f) \text{ for all }, S_eS_f \in J..$$

Example 1.14. Given IVIFG $G = (A_1, A_2)$ as shown in **Figure 1**.
From the given IVIFG we have

$$\sigma_{A_1}(v_1) = \left[\sigma^-_{A_1}(v_1), \sigma^+_{A_1}(v_1)\right] = [0.3,0.6]$$
$$\sigma_{A_1}(v_2) = \left[\sigma^-_{A_1}(v_2), \sigma^+_{A_1}(v_2)\right] = [0.2,0.7]$$
$$\sigma_{A_1}(v_3) = \left[\sigma^-_{A_1}(v_3), \sigma^+_{A_1}(v_3)\right] = [0.1,0.3]$$
$$\sigma_{A_1}(v_4) = \left[\sigma^-_{A_1}(v_4), \sigma^+_{A_1}(v_4)\right] = [0.3,0.4]$$

$$\gamma_{A_1}(v_1) = \left[\gamma^-_{A_1}(v_1), \gamma^+_{A_1}(v_1)\right] = [0.1,0.4]$$
$$\gamma_{A_1}(v_2) = \left[\gamma^-_{A_1}(v_2), \gamma^+_{A_1}(v_2)\right] = [0.1,0.2]$$
$$\gamma_{A_1}(v_3) = \left[\gamma^-_{A_1}(v_3), \gamma^+_{A_1}(v_3)\right] = [0.4,0.5]$$
$$\gamma_{A_1}(v_4) = \left[\gamma^-_{A_1}(v_4), \gamma^+_{A_1}(v_4)\right] = [0.4,0.5]$$

$$\sigma_{B_1}(v_1v_2) = \left[\sigma^-_{B_1}(v_1v_2), \sigma^+_{B_1}(v_1v_2)\right] = [0.2,0.5]$$
$$\sigma_{B_1}(v_2v_3) = \left[\sigma^-_{B_1}(v_2v_3), \sigma^+_{B_1}(v_2v_3)\right] = [0.1,0.2]$$
$$\sigma_{B_1}(v_3v_4) = \left[\sigma^-_{B_1}(v_3v_4), \sigma^+_{B_1}(v_3v_4)\right] = [0.1,0.1]$$
$$\sigma_{B_1}(v_4v_1) = \left[\sigma^-_{B_1}(v_4v_1), \sigma^+_{B_1}(v_4v_1)\right] = [0.2,0.4]$$

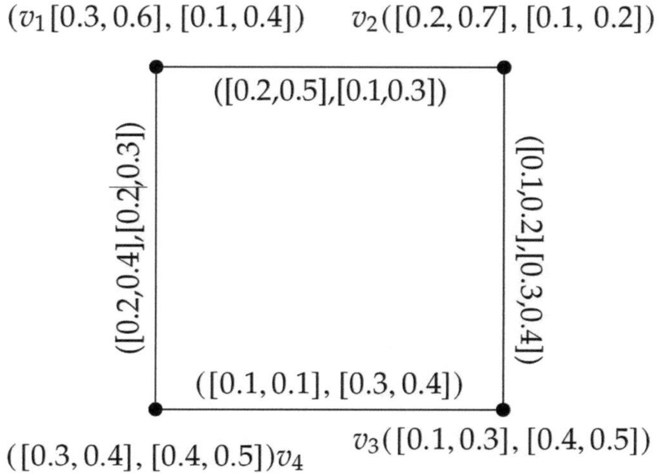

Figure 1.
IVIFG G.

$$\gamma_{B_1}(v_1v_2) = \left[\gamma_{B_1}^{-}(v_1v_2), \gamma_{B_1}^{+}(v_1v_2)\right] = [0.1,0.3]$$
$$\gamma_{B_1}(v_2v_3) = \left[\gamma_{B_1}^{-}(v_2v_3), \gamma_{B_1}^{+}(v_2v_3)\right] = [0.3,0.4]$$
$$\gamma_{B_1}(v_3v_4) = \left[\gamma_{B_1}^{-}(v_3v_4), \gamma_{B_1}^{+}(v_3v_4)\right] = [0.3,0.4]$$
$$\gamma_{B_1}(v_4v_1) = \left[\gamma_{B_1}^{-}(v_4v_1), \gamma_{B_1}^{+}(v_4v_1)\right] = [0.2,0.3]$$

To find IVIFLG $L(G) = (H,J)$ of I such that

$$H = \{v_1v_2 = S_{e_1}, v_2v_3 = S_{e_2}, v_3v_4 = S_{e_3}, v_4v_1 = S_{e_4}\} \text{ and}$$
$$J = \{S_{e_1}S_{e_2}, S_{e_2}S_{e_3}, S_{e_2 3}S_{e_4}, S_{e_4}S_{e_1}\}.$$

Now, consider $A_2 = \left[\sigma_{A_2}^{-}, \sigma_{A_2}^{+}\right]$ and $B_2 = \left[\sigma_{B_2}^{-}, \sigma_{B_2}^{+}\right]$ are IVFS of H and J respectively. Then we have

$$\sigma_{A_2}(S_{e_1}) = \left[\sigma_{B_1}^{-}(e_1), \sigma_{B_1}^{+}(e_1)\right] = [0.2,0.5]$$
$$\sigma_{A_2}(S_{e_2}) = \left[\sigma_{B_1}^{-}(e_2), \sigma_{B_1}^{+}(e_2)\right] = [0.1,0.2]$$
$$\sigma_{A_2}(S_{e_3}) = \left[\sigma_{B_1}^{-}(e_3), \sigma_{B_1}^{+}(e_3)\right] = [0.1,0.1]$$
$$\sigma_{A_2}(S_{e_4}) = \left[\sigma_{B_1}^{-}(e_4), \sigma_{B_1}^{+}(e_4)\right] = [0.2,0.4]$$

$$\gamma_{A_2}(S_{e_1}) = \left[\gamma_{B_1}^{-}(e_1), \gamma_{B_1}^{+}(e_1)\right] = [0.1,0.3]$$
$$\gamma_{A_2}(S_{e_2}) = \left[\gamma_{B_1}^{-}(e_2), \gamma_{B_1}^{+}(e_2)\right] = [0.3,0.4]$$
$$\gamma_{A_2}(S_{e_3}) = \left[\gamma_{B_1}^{-}(e_3), \gamma_{B_1}^{+}(e_3)\right] = [0.3,0.4]$$
$$\gamma_{A_2}(S_{e_4}) = \left[\gamma_{B_1}^{-}(e_4), \gamma_{B_1}^{+}(e_4)\right] = [0.2,0.3]$$

$$\sigma_{B_2}(S_{e_1}S_{e_2}) = \left[\sigma_{B_1}^{-}(e_1)\wedge\sigma_{B_1}^{-}(e_2), \sigma_{B_1}^{+}(e_1)\wedge\sigma_{B_1}^{+}(e_2)\right] = [0.1,0.2]$$
$$\sigma_{B_2}(S_{e_2}S_{e_3}) = \left[\sigma_{B_1}^{-}(e_2)\wedge\sigma_{B_1}^{-}(e_3), \sigma_{B_1}^{+}(e_2)\wedge\sigma_{B_1}^{+}(e_3)\right] = [0.1,0.1]$$
$$\sigma_{B_2}(S_{e_3}S_{e_4}) = \left[\sigma_{B_1}^{-}(e_3)\wedge\sigma_{B_1}^{-}(e_4), \sigma_{B_1}^{+}(e_3)\wedge\sigma_{B_1}^{+}(e_4)\right] = [0.1,0.1]$$
$$\sigma_{B_2}(S_{e_2}S_{e_3}) = \left[\sigma_{B_1}^{-}(e_4)\wedge\sigma_{B_1}^{-}(e_1), \sigma_{B_1}^{+}(e_4)\wedge\sigma_{B_1}^{+}(e_1)\right] = [0.2,0.4]$$

$$\gamma_{B_2}(S_{e_1}S_{e_2}) = \left[\gamma_{B_1}^{-}(e_1)\vee\gamma_{B_1}^{-}(e_2), \gamma_{B_1}^{+}(e_1)\vee\gamma_{B_1}^{+}(e_2)\right] = [0.3,0.4]$$
$$\gamma_{B_2}(S_{e_2}S_{e_3}) = \left[\gamma_{B_1}^{-}(e_2)\vee\gamma_{B_1}^{-}(e_3), \gamma_{B_1}^{+}(e_2)\vee\gamma_{B_1}^{+}(e_3)\right] = [0.3,0.4]$$
$$\gamma_{B_2}(S_{e_3}S_{e_4}) = \left[\gamma_{B_1}^{-}(e_3)\vee\gamma_{B_1}^{-}(e_4), \gamma_{B_1}^{+}(e_3)\vee\gamma_{B_1}^{+}(e_4)\right] = [0.3,0.4]$$
$$\gamma_{B_2}(S_{e_2}S_{e_3}) = \left[\gamma_{B_1}^{-}(e_4)\vee\gamma_{B_1}^{-}(e_1), \gamma_{B_1}^{+}(e_4)\vee\gamma_{B_1}^{+}(e_1)\right] = 0.2,0.3]$$

Then L(G) of IVIFG G is shown in **Figure 2**.

Proposition 1.15. $L(G) = (A_2, B_2)$ is IVIFLG corresponding to IVIFG $G = (A_1, B_1)$.

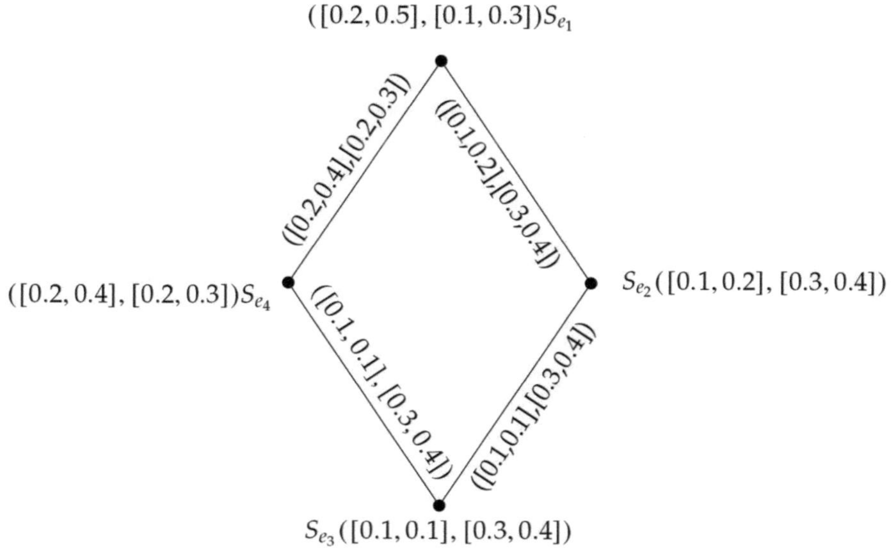

Figure 2.
IVIFLG of G.

Definition 1.16. A homomorphism mapping $\psi : G_1 \to G_2$ of two IVIFG $G_1 = (M_1, N_1)$ and $G_2 = (M_2, N_2)$ $\psi : V_1 \to V_2$ is defined as

i. $\sigma^-_{M_1}(v_i) \le \sigma^-_{M_2}(\psi(v_i))$, $\quad \sigma^+_{M_1}(v_i) \le \sigma^+_{M_2}(\psi(v_i))$

$\gamma^-_{M_1}(v_i) \le \gamma^-_{M_2}(\psi(v_i))$, $\quad \gamma^+_{M_1}(v_i) \le \gamma^+_{M_2}(\psi(v_i))$ for all $v_i \in V_1$.

ii. $\sigma^-_{N_1}(v_iv_j) \le \sigma^-_{N_2}(\psi(v_i)\psi(v_j))$, $\quad \sigma^+_{N_1}(v_iv_j) \le \sigma^+_{N_2}(\psi(v_i)\psi(v_j))$

$\gamma^-_{N_1}(v_iv_j) \le \gamma^-_{N_2}(\psi(v_i)\psi(v_j))$, $\quad \gamma^+_{N_1}(v_iv_j) \le \gamma^+_{N_2}(\psi(v_i)\psi(v_j))$ for all $v_iv_j \in E_1$.

Definition 1.17. A bijective homomorphism $\psi : G_1 \to G_2$ of IVIFG is said to be a weak vertex isomorphism, if

$$\sigma_{M_1}(v_i) = \left[\sigma^-_{M_1}(v_i), \sigma^+_{M_1}(v_i)\right] = \left[\sigma^-_{M_2}(\psi(v_i)), \sigma^+_{M_2}(\psi(v_i))\right]$$
$$\gamma_{N_1}(v_i) = \left[\gamma^-_{N_1}(v_i), \gamma^+_{N_1}(v_i)\right] = \left[\gamma^-_{N_2}(\psi(v_i)), \gamma^+_{N_2}(\psi(v_i))\right], \quad \forall v_i \in V_1.$$

A bijective homomorphism $\psi : G_1 \to G_2$ of IVIFG is said to be a weak line isomorphism if

$$\sigma_{B_1}(v_iv_j) = \left[\sigma^-_{B_1}(v_iv_j), \sigma^+_{B_1}(v_iv_j)\right] = \left[\sigma^-_{B_2}(\psi(v_i)\psi(v_j)), \sigma^+_{B_2}(\psi(v_i)\psi(v_j))\right],$$
$$\gamma_{B_1}(v_iv_j) = \left[\gamma^-_{B_1}(v_iv_j), \gamma^+_{B_1}(v_iv_j)\right] = \left[\gamma^-_{B_2}(\psi(v_i)\psi(v_j)), \gamma^+_{B_2}(\psi(v_i)\psi(v_j))\right] \quad \forall v_iv_j \in E_1.$$

If $\psi : G_1 \to G_2$ is an isomorphism that holds Definition 1.17, then ψ is called a weak isomorphism of IVIFGs G_1 and G_2.

Proposition 1.18. The IVIFLG $L(G)$ is connected graph if and only if its corresponding IVFG G is connected graph.

Proof: Assume that $L(G)$ is a connected IVIFLG of the IVIFG G. First, We want to show that necessary condition. Lets say G is disconnected IVIFG. Then there are at least two nodes of graph G which are not joined by path, say v_i and v_j. If we take one edge e in the first component of the edge set of G, then it doesn't have any edges which adjacent to edge e in other components. So that, the IVIFLG of graph G is disconnected and contradicts our assumption. Therefore, the IVIFG G must be connected. On the other hand, assume that IVIFG G is connected graph. Then, there is a path between each pair of nodes. This implies, edges which are adjacent in graph G are adjacent nodes in IVIFLG. As a result, every pair of nodes in IVIFLG of G are linked by a path. Therefore, the proof finished.

Proposition 1.19. An Interval valued line graph of star graph $K_{1,n}$ is a complete Interval valued graph K_n with $n-$vertices.

Proof: Consider the vertex $v \in V(K_{1,n})$ that adjacent to all other vertices $u_i \in V(K_{1,n})$ for $i = 1, 2\cdots, n$. Now, all the vertices in IVIFLG of $K_{1,n}$ are adjacent. This means, IVIFLG of $K_{1,n}$ is a complete graph.

Example 1.20. Suppose that the IVIFG $K_{1,3}$ with $V = \{v, v_1, v_2, v_3\}$ and $E = \{vv_1, vv_2, vv_3$ where

$$v = ([0.3,0.5],[0.1,0.4]),\ v_1 = ([0.3,0.4],[0.2,0.5])$$
$$v_2 = ([0.5,0.8],[0.1,0.2]),\ v_3 = ([0.1,0.3],[0.5,0.7])$$
$$e_1 = vv_1 = ([0.2,0.3],[0.3,0.5]),\ e_2 = vv_2 = ([0.2,0.5],[0.0,0.3])$$
$$e_3 = vv_3 = ([0.1,0.2],[0.3,0.6]).$$

Then by definition of IVIFLG, the vertex sets of $L(K_{1,3})$ is $V = \{S_{e_1}, S_{e_2}, S_{e_3}\}$ and $\{S_{e_1}S_{e_2}, S_{e_1}S_{e_3}, S_{e_2}S_{e_3}$ edge sets where

$$S_{e_1} = ([0.2,0.3],[0.3,0.5]), \qquad S_{e_2} = ([0.2,0.5],[0.0,0.3]),$$
$$S_{e_3} = ([0.1,0.2],[0.2,0.6]), \qquad S_{e_1}S_{e_2} = ([0.2,0.3],[0.3,0.5]),$$
$$S_{e_1}S_{e_3} = ([0.2,0.3],[0.3,0.5]), \qquad S_{e_2}S_{e_3} = ([0.1,0.2],[0.2,0.6]).$$

Here $L(K_{1,3})$ is complete graph K_3 (**Figure 3**).

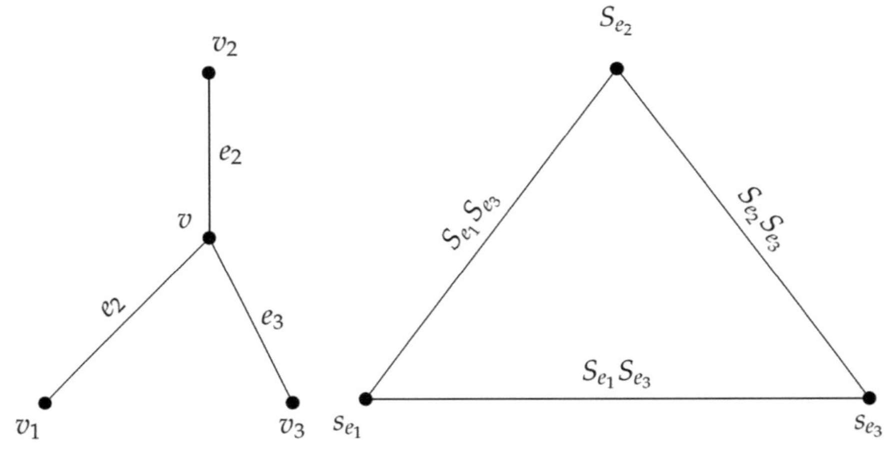

Figure 3.
Graphs of $K_{1,3}$ and $L(K_{1,3})$.

Proposition 1.21. Let $L(G)$ be IVIFLG of IVIFG of G. Then $L(G^*)$ is a line graph of G^* where $G^* = (V, E)$ with underlying set V.

Proof: Given $G = (A_1, B_1)$ is IVIFG of G^* and $L(G) = (A_2, B_2)$ is IVIFLG of $L(G^*)$. Then

$$\sigma_{A_2}(S_e) = \left[\sigma^-_{A_2}(S_e), \sigma^+_{A_2}(S_e)\right] = \left[\sigma^-_{B_1}(e), \sigma^+_{B_1}(e)\right],$$

$$\gamma_{A_2}(S_e) = \left[\gamma^-_{A_2}(S_e), \gamma^+_{A_2}(S_e)\right] = \left[\gamma^-_{B_1}(e), \gamma^+_{B_1}(e)\right] \ \forall e \in E.$$

This implies, $S_e \in H = \{\{e\} \cup \{u_e, v_e\} : e \in E, u_e, v_e \in V \& e = u_e v_e\}$ if and only if $e \in E$.

$$\sigma_{B_2}(S_e S_f) = \left[\sigma^-_{B_2}(S_e S_f), \sigma^+_{B_2}(S_e S_f)\right] = \left[\sigma^-_{B_1}(e) \wedge \sigma^-_{B_1}(f), \sigma^+_{B_1}(e) \wedge \sigma^+_{B_1}(f)\right]$$

$$\gamma_{B_2}(S_e S_f) = \left[\gamma^-_{B_2}(S_e S_f), \gamma^+_{B_2}(S_e S_f)\right] = \left[\gamma^-_{B_1}(e) \vee \gamma^-_{B_1}(f), \gamma^+_{B_1}(e) \vee \gamma^+_{B_1}(f)\right]$$

$$\forall S_e S_f \in J,$$

where $J = \{S_e S_f \mid S_e \cap S_f \notin \varnothing, e, f \in E \& e \notin f\}$. Hence, $L(G^*)$ is a line graph of G^*.

Proposition 1.22. Let $L(G) = (A_2, B_2)$ be IVIFLG of $L(G^*)$. Then $L(G)$ is also IVIFLG of some IVIFG $G = (A_1, B_1)$ iff

i. $\sigma_{B_2}(S_e S_f) = \left[\sigma^-_{B_2}(S_e S_f), \sigma^+_{B_2}(S_e S_f)\right] = \left[\sigma^-_{A_2}(S_e) \wedge \sigma^-_{A_2}(S_f), \sigma^+_{A_2}(S_e) \wedge \sigma^+_{A_2}(S_f)\right],$

ii. $\gamma_{B_2}(S_e S_f) = \left[\gamma^-_{B_2}(S_e S_f), \gamma^+_{B_2}(S_e S_f)\right] = \left[\gamma^-_{A_2}(S_e) \vee \gamma^-_{A_2}(S_f), \gamma^+_{A_2}(S_e) \vee \gamma^+_{A_2}(S_f)\right]$
$\forall S_e, S_f \in H, S_e S_f \in J.$

Proof: Suppose both conditions (i) and (ii) are satisfied. i.e., $\sigma^-_{B_2}(S_e S_f) = \sigma^-_{A_2}(S_e) \wedge \sigma^-_{A_2}(S_f)$, $\sigma^+_{B_2}(S_e S_f) = \sigma^+_{A_2}(S_e) \wedge \sigma^+_{A_2}(S_f)$, $\gamma^-_{B_2}(S_e S_f) = \gamma^-_{A_2}(S_e) \vee \gamma^-_{A_2}(S_f)$ and $\gamma^+_{B_2}(S_e S_f) = \gamma^+_{A_2}(S_e) \vee \gamma^+_{A_2}(S_f)$ for all $S_e S_f \in W$. For every $e \in E$ we define $\sigma^-_{A_2}(S_e) = \sigma^-_{A_1}(e)$, $\sigma^+_{A_2}(S_e) = \sigma^+_{A_1}(e)$, $\gamma^-_{A_2}(S_e) = \gamma^-_{A_1}(e)$ and $\gamma^+_{A_2}(S_e) = \gamma^+_{A_1}(e)$. Then

$$\begin{aligned}
\sigma^-_{B_2}(S_e S_f) &= \left[\sigma^-_{B_2}(S_e S_f), \sigma^+_{B_2}(S_e S_f)\right] \\
&= \left[\sigma^-_{A_2}(S_e) \wedge \sigma^-_{A_2}(S_f), \sigma^+_{A_2}(S_e) \wedge \sigma^+_{A_2}(S_f)\right] \\
&= \left[\sigma^-_{B_1}(e) \wedge \sigma^-_{B_1}(f), \sigma^+_{B_1}(e) \wedge \sigma^+_{B_1}(f)\right]. \\
\gamma^-_{B_2}(S_e S_f) &= \left[\gamma^-_{B_2}(S_e S_f), \gamma^+_{B_2}(S_e S_f)\right] \\
&= \left[\gamma^-_{A_2}(S_e) \vee \gamma^-_{A_2}(S_f), \gamma^+_{A_2}(S_e) \vee \gamma^+_{A_2}(S_f)\right] \\
&= \left[\gamma^-_{B_1}(e) \vee \gamma^-_{B_1}(f), \gamma^+_{B_1}(e) \vee \gamma^+_{B_1}(f)\right].
\end{aligned}$$

We know that IVIFS $A_1 = \left(\left[\sigma^-_{A_1}, \sigma^+_{A_1}\right], \left[\gamma^-_{A_1}, \gamma^+_{A_1}\right]\right)$ yields the properties

$$\sigma_{B_1}^-(v_iv_j) \leq \sigma_{A_1}^-(v_i) \wedge \sigma_{A_1}^-(v_j)$$
$$\sigma_{B_1}^+(v_iv_j) \leq \sigma_{A_1}^+(v_i) \wedge \sigma_{A_1}^+(v_j)$$
$$\gamma_{B_1}^-(v_iv_j) \leq \gamma_{A_1}^-(v_i) \vee \gamma_{A_1}^-(v_j)$$
$$\gamma_{B_1}^+(v_iv_j) \leq \gamma_{A_1}^+(v_i) \vee \gamma_{A_1}^+(v_j)$$

will suffice. From definition of IVIFLG the converse of this statement is well known.

Proposition 1.23. An IVIFLG is always a strong IVIFG.

Proof: It is straightforward from the definition, therefore it is omitted.

Proposition 1.24. Let G_1 and G_2 IVIFGs of G_1^* and G_2^* respectively. If the mapping $\psi : G_1 \to G_2$ is a weak isomorphism, then $\psi : G_1^* \to G_2^*$ is isomorphism map.

Proof: Suppose $\psi : G_1 \to G_2$ is a weak isomorphism. Then

$$v \in V_1 \Leftrightarrow \psi(v) \in V_2 \text{ and}$$

$$uv \in E_1 \Leftrightarrow \psi(u)\psi(v) \in E_2.$$

Hence the proof.

Theorem 1.25. Let $G^* = (V, E)$ is connected graph and consider that $L(G) = (A_2, B_2)$ is IVIFLG corresponding to IVIFG $G = (A_1, B_1)$. The,

1. there exists a map $\psi : G \to L(G)$ which is a weak isomorphism if and only if G^* is a cyclic graph with

   $$\sigma_{A_1}(v) = \left[\sigma_{A_1}^-(v), \sigma_{A_1}^+(v)\right] = \left[\sigma_{B_1}^-(e), \sigma_{B_1}^+(e)\right],$$

   $$\gamma_{A_1}(v) = \left[\gamma_{A_1}^-(v), \gamma_{A_1}^+(v)\right] = \left[\gamma_{B_1}^-(e), \gamma_{B_1}^+(e)\right],$$

   such that $A_1 = \left(\left[\sigma_{A_1}^-, \sigma_{A_1}^+\right], \left[\gamma_{A_1}^-, \gamma_{A_1}^+\right]\right) \& B_1 = \left(\left[\sigma_{B_1}^-, \sigma_{B_1}^+\right], \left[\gamma_{B_1}^-, \gamma_{B_1}^+\right]\right)$, $\forall v \in V, e \in E$.

2. The map ψ is isomorphism if $\psi : G \to L(G)$ is a weak isomorphism.

Proof: Consider $\psi : G \to L(G)$ is a weak isomorphism. Then we have

$$\sigma_{A_1}(v_i) = \left[\sigma_{A_1}^-(v_i), \sigma_{A_1}^+(v_i)\right] = \left[\sigma_{A_2}^-(\psi(v_i)), \sigma_{A_2}^+(\psi(v_i))\right]$$
$$\gamma_{B_1}(v_i) = \left[\gamma_{B_1}^-(v_i), \gamma_{B_1}^+(v_i)\right] = \left[\gamma_{B_2}^-(\psi(v_i)), \gamma_{B_2}^+(\psi(v_i))\right]$$
$$\forall v_i \in V.$$

$$\sigma_{B_1}(v_iv_j) = \left[\sigma_{B_1}^-(v_iv_j), \sigma_{B_1}^+(v_iv_j)\right] = \left[\sigma_{B_2}^-(\psi(v_i)\psi(v_j)), \sigma_{B_2}^+(\psi(v_i)\psi(v_j))\right]$$

$$\gamma_{B_1}(v_iv_j) = \left[\gamma_{B_1}^-(v_iv_j), \gamma_{B_1}^+(v_iv_j)\right] = \left[\gamma_{B_2}^-(\psi(v_i)\psi(v_j)), \gamma_{B_2}^+(\psi(v_i)\psi(v_j))\right] \ \forall v_iv_j \in E.$$

This follows that $G^* = (V, E)$ is a cyclic from Proposition 1.24.

Now let $v_1v_2v_3 \cdots v_nv_1$ be a cycle of G^* where vertices set $V = \{v_1, v_2, \cdots, v_n\}$ and edges set $E = \{v_1v_2, v_2v_3, \cdots, v_nv_1\}$. Then we have IVIFS

$$\sigma_{A_1}(v_i) = \left[\sigma^-_{A_1}(v_i), \sigma^+_{A_1}(v_i)\right] = \left[t^-_i, t^+_i\right]$$

$$\gamma_{A_1}(v_i) = \left[\gamma^-_{A_1}(v_i), \gamma^+_{A_1}(v_i)\right] = \left[f^-_i, f^+_i\right]$$

and

$$\sigma_{B_1}(v_i v_{i+1}) = \left[\sigma^-_{B_1}(v_i v_{i+1}), \sigma^+_{B_1}(v_i v_{i+1})\right] = \left[\iota^-_i, \iota^+_i\right]$$

$$\gamma_{B_1}(v_i v_{i+1}) = \left[\gamma^-_{B_1}(v_i v_{i+1}), \gamma^+_{B_1}(v_i v_{i+1})\right] = \left[q^-_i, q^+_i\right],$$

where $i = 1, 2, \cdots, n$ and $v_{n+1} = v_1$. Thus, for $t^-_1 = t^-_{n+1}, t^+_1 = t^+_{n+1}, f^-_1 = f^-_{n+1}, f^+_1 = f^-_{n+1}$

$$\begin{aligned} &\iota^-_i \leq t^-_i \wedge t^-_{i+1}, \\ &\iota^+_i \leq t^+_i \wedge t^+_{i+1}, \\ &q^-_i \leq f^-_i \vee f^-_{i+1} \\ &q^+_i \leq f^+_i \vee f^+_{i+1}. \end{aligned} \tag{1}$$

Now

$$H = \{S_{e_i} : i = 1, 2, , \cdots, n\} \text{ and } J = \left\{S_{e_i} S_{e_{i+1}} : i = 1, 2, , \cdots, n-1 \ \right\}.$$

And also,

$$\begin{aligned} \sigma_{A_2}(S_{e_i}) &= \left[\sigma^-_{A_2}(S_{e_i}), \sigma^+_{A_2}(S_{e_i})\right] \\ &= \left[\sigma^-_{B_1}(e_i), \sigma^+_{B_1}(e_i)\right] \\ &= \left[\sigma^-_{B_1}(v_i v_{i+1}), \sigma^+_{B_1}(v_i v_{i+1})\right] \\ &= \left[\iota^-_i, \iota^+_i\right] \\ \gamma_{A_2}(S_{e_i}) &= \left[\gamma^-_{A_2}(S_{e_i}), \gamma^+_{A_2}(S_{e_i})\right] \\ &= \left[\gamma^-_{B_1}(e_i), \gamma^+_{B_1}(e_i)\right] \\ &= \left[\gamma^-_{B_1}(v_i v_{i+1}), \gamma^+_{B_1}(v_i v_{i+1})\right] \\ &= \left[q^-_i, q^+_i\right] \\ \sigma^+_{B_2}\left(S_{e_i} S_{e_{i+1}}\right) &= min\left\{\sigma^+_{B_1}(e), \sigma^+_{B_1}(e_{i+1})\right\} \\ &= min\left\{\sigma^+_{B_1}(v_i v_{i+1}), \sigma^+_{B_1}(v_{i+1} v_{i+2})\right\} \\ &= min\left\{\iota^+_i, \iota^+_{i+1}\right\} \\ \sigma^-_{B_2}\left(S_{e_i} S_{e_{i+1}}\right) &= min\left\{\ \sigma^-_{B_1}(e), \sigma^-_{B_1}(e_{i+1})\right\} \\ &= min\left\{\ \sigma^-_{B_1}(v_i v_{i+1}), \sigma^-_{B_1}(v_{i+1} v_{i+2})\right\} \\ &= min\left\{\iota^-_i, \iota^-_{i+1}\right\} \end{aligned}$$

$$\begin{aligned}\gamma^{+}_{B_2}\left(S_{e_i}S_{e_{i+1}}\right) &= max\left\{\gamma^{+}_{B_1}(e),\gamma^{+}_{B_1}(e_{i+1})\right\}\\ &= max\left\{\gamma^{+}_{B_1}(v_iv_{i+1}),\gamma^{+}_{B_1}(v_{i+1}v_{i+2})\right\}\\ &= max\left\{q^{+}_{i},q^{+}_{i+1}\right\}\\ \gamma^{-}_{B_2}\left(S_{e_i}S_{e_{i+1}}\right) &= max\left\{\gamma^{-}_{B_1}(e),\gamma^{-}_{B_1}(e_{i+1})\right\}\\ &= max\left\{\gamma^{-}_{B_1}(v_iv_{i+1}),\gamma^{-}_{B_1}(v_{i+1}v_{i+2})\right\}\\ &= max\left\{q^{-}_{i},q^{-}_{i+1}\right\}\end{aligned}$$

where $v_{n+1}=v_1, v_{n+2}=v_2, \iota^{+}_{1}=\iota^{+}_{n+1}, \iota^{-}_{1}=\iota^{-}_{n+1}, q^{+}_{n+1}=\iota^{+}_{1},, q^{-}_{n+1}=q^{-}_{1}$, and $i=1,2,\cdots,n.\psi: V\to H$ is bijective map since $\psi: G^{*}\to L(G^{*})$ is isomorphism. And also, ψ preserves adjacency. So that ψ persuades an alternative τ of $\{1,2,\cdots,n\}$ which $\psi(v_i)=S_{e_{\tau(i)}}$ and for $e_i=v_iv_{i+1}$ then $\psi(v_i)\psi(v_{i+1})=S_{e_{\tau(i)}}S_{e_{\tau(i+1)}}, i=1,2,\cdots,n-1$. Now

$$\begin{aligned}t^{-}_{i} &= \sigma^{-}_{A_1}(v_i)\leq\sigma^{-}_{A_2}(\psi(v_i))=\sigma^{-}_{A_2}\left(S_{e_{\tau(i)}}\right)=\iota^{-}_{\tau(i)},\\ t^{+}_{i} &= \sigma^{+}_{A_1}(v_i)\leq\sigma^{+}_{A_2}(\psi(v_i))=\sigma^{+}_{A_2}\left(S_{e_{\tau(i)}}\right)=\iota^{+}_{\tau(i)},\\ f^{-}_{i} &= \gamma^{-}_{A_1}(v_i)\leq\gamma^{-}_{A_2}(\psi(v_i))=\gamma^{-}_{A_2}\left(S_{e_{\tau(i)}}\right)=q^{-}_{\tau(i)},\\ f^{+}_{i} &= \gamma^{+}_{A_1}(v_i)\leq\gamma^{+}_{A_2}(\psi(v_i))=\gamma^{+}_{A_2}\left(S_{e_{\tau(i)}}\right)=q^{+}_{\tau(i)}.\end{aligned}$$

And let $e_i=v_iv_{i+1}$,

$$\begin{aligned}\iota^{-}_{i} &= \sigma^{-}_{B_1}(v_iv_{i+1})\leq\sigma^{-}_{B_2}\left(\psi(v_i)\psi(v_{i+1})=\sigma^{-}_{B_2}\left(S_{e_{\tau(i)}}S_{e_{\tau(i+1}}\right)\right)\\ &= min\left\{\sigma^{-}_{B_1}(e_{\tau(i)}),\sigma^{-}_{B_1}(e_{\tau(i+1)})\right\}=min\left\{\iota^{-}_{\tau(i)},\iota^{-}_{\tau(i+1)}\right\}\\ \iota^{+}_{i} &= \sigma^{+}_{B_1}(v_iv_{i+1})\leq\sigma^{+}_{B_2}\left(\psi(v_i)\psi(v_{i+1})=\sigma^{+}_{B_2}\left(S_{e_{\tau(i)}}S_{e_{\tau(i+1}}\right)\right)\\ &= min\left\{\sigma^{+}_{B_1}(e_{\tau(i)}),\sigma^{+}_{B_1}(e_{\tau(i+1)})\right\}=min\left\{\iota^{+}_{\tau(i)},\iota^{+}_{\tau(i+1)}\right\}\\ q^{-}_{i} &= \gamma^{-}_{B_1}(v_iv_{i+1})\leq\gamma^{-}_{B_2}\left(\psi(v_i)\psi(v_{i+1})=\gamma^{-}_{B_2}\left(S_{e_{\tau(i)}}S_{e_{\tau(i+1}}\right)\right)\\ &= max\left\{\gamma^{-}_{B_1}(e_{\tau(i)}),\gamma^{-}_{B_1}(e_{\tau(i+1)})\right\}=max\left\{q^{-}_{\tau(i)},q^{-}_{\tau(i+1)}\right\}\\ q^{+}_{i} &= \gamma^{+}_{B_1}(v_iv_{i+1})\leq\gamma^{+}_{B_2}\left(\psi(v_i)\psi(v_{i+1})=\gamma^{+}_{B_2}\left(S_{e_{\tau(i)}}S_{e_{\tau(i+1}}\right)\right)\\ &= max\left\{\gamma^{+}_{B_1}(e_{\tau(i)}),\gamma^{+}_{B_1}(e_{\tau(i+1)})\right\}\\ &= max\left\{q^{+}_{\tau(i)},q^{+}_{\tau(i+1)}\right\}\ for\ i=1,2,\cdots,n.\end{aligned}$$

Which implies,

$$\begin{aligned}t^{-}_{i}\leq\iota^{-}_{\tau(i)},\quad t^{+}_{i}\leq\iota^{+}_{\tau(i)}\\ f^{-}_{i}\leq q^{-}_{\tau(i)},\quad f^{+}_{i}\leq q^{+}_{\tau(i)}\end{aligned}\tag{2}$$

and

$$\begin{aligned} \iota_i^- &\le min\left\{\iota_{\tau(i)}^-, \iota_{\tau(i+1)}^-\right\}, \quad \iota_i^+ \le min\left\{\iota_{\tau(i)}^+, \iota_{\tau(i+1)}^+\right\} \\ q_i^- &\le max\left\{q_{\tau(i)}^-, q_{\tau(i+1)}^-\right\}, \quad q_i^+ \le max\left\{q_{\tau(i)}^+, q_{\tau(i+1)}^+\right\}. \end{aligned} \tag{3}$$

Thus from the above equations, we obtain $\iota_i^- \le \iota_{\tau(i)}^-, \iota_i^+ \le \iota_{\tau(i)}^+, q_i^- \le q_{\tau(i)}^-$ and $q_i^+ \le q_{\tau(i)}^+$. and also $\iota_{\tau(i)}^- \le \iota_{\tau(\tau(i))}^-, \iota_{\tau(i)}^+ \le \iota_{\tau(\tau(i))}^+, q_{\tau(i)}^- \le q_{\tau(\tau(i))}^-$ and $q_{\tau(i)}^+ \le q_{\tau(\tau(i))}^+$. By proceeding this process, we get

$$\begin{aligned} \iota_i^- &\le \iota_{\tau(i)}^- \le \cdots \le \iota_{\tau^k(i)}^- \le \iota_i^- \\ \iota_i^+ &\le \iota_{\tau(i)}^+ \le \cdots \le \iota_{\tau^k(i)}^+ \le \iota_i^+ \\ q_i^- &\le q_{\tau(i)}^- \le \cdots \le q_{\tau^k(i)}^- \le q_i^- \\ q_i^+ &\le q_{\tau(i)}^+ \le \cdots \le q_{\tau^k(i)}^+ \le q_i^+ \end{aligned}$$

where τ^{k+1} is the identity function. It follows $\iota_{\tau(i)}^- = \iota_{\tau(\tau(i))}^-, \iota_{\tau(i)}^+ = \iota_{\tau(\tau(i))}^+, q_{\tau(i)}^- = q_{\tau(\tau(i))}^-$ and $q_{\tau(i)}^+ = q_{\tau(\tau(i))}^+$. Again, from Eq. (3), we get

$$\begin{aligned} \iota_i^- &\le \iota_{\tau(i+1)}^- = \iota_{i+1}^-, \quad \iota_i^+ \le \iota_{\tau(i+1)}^+ = \iota_{i+1}^+ \\ q_i^- &\le q_{\tau(i+1)}^- = q_{i+1}^-, q_i^+ \le q_{\tau(i+1)}^+ = q_{i+1}^-. \end{aligned}$$

This implies for all $i = 1, 2, \cdots, n$, $\iota_i^- = \iota_1^-, \iota_i^+ = \iota_1^+, q_i^- = q_1^-$ and $q_i^+ = q{+}_1$. Thus, from Eqs. (1) and (2) we obtain

$$\begin{aligned} \iota_1^- &= \cdots = \iota_n^- = t_1^- = \cdots = t_n^- \\ \iota_1^+ &= \cdots = \iota_n^+ = t_1^+ = \cdots = t_n^+ \\ q_1^- &= \cdots = q_n^- = f_1^- = \cdots = f_n^- \\ q_1^+ &= \cdots = q_n^+ = f_1^+ = \cdots = f_n^+. \end{aligned}$$

As a result, the proof.

Theorem 1.26. Let G be connected simple IVIFG, then IVIFLG of G is a path graph if and only if G is path graph.

Proof: Suppose that G is a path IVIFG with $|V(G)| = k$. Thus, G is a path P_k with length k and $|E(G)| = k - 1$. Since the vertices set of IVIFLG $L(G)$ is an edge sets of G, clearly $L(G)$ is a path with $|V(L(G))| = k - 1$ graph and $|E(L(G))| = k - 2$. Implies that $L(G)$ is a path graph. On the other hand, assume $L(G)$ is a path. Then every degree of vertex $v_i \in G$ is can't be greater than two. If there is a vertex $v_i \in G$ is greater than two, then an edge e which incident to $v_i \in G$ would form a complete sub-graph of IVIFLG $L(G)$ of more than two vertices. As a result, the IVIFG G must be either path graph or cyclic. But, G can't be the cyclic graph since a line graph of the cyclic graph is the cyclic graph. The proof is finished.

3. Conclusion

In this chapter, we introduced interval-valued intuitionistic fuzzy line graphs (IVIFLG) and investigated their results. In addition, we developed many theorems, and propositions related to IVIFLG with proof. Moreover, some remarkable properties of isomorphic properties, strong IVIFLG, and complete IVIFLG have been investigated, and the proposed concepts are illustrated with the examples.

Acknowledgements

The authors do thankful to the editor for giving an opportunity to submit our book chapter.

Competing interest

The authors declared that they have no competing interests.

Author details

Venkata Naga Srinivasa Rao Repalle[1*†], Keneni Abera Tola[2†] and Maamo Abebe Ashebo[1†]

1 Wollega University, Nekemte, Ethiopia

2 Dambi Dollo University, Dambi Dollo, Ethiopia

*Address all correspondence to: rvnrepalle@gmail.com

† These authors contributed equally.

References

[1] Zadeh AL. Information and control. Fuzzy Sets. 1965;**8**(3):338-353

[2] Kaufmann A. Introduction Theory of Fuzzy Sets. New York: Academic Press; 1975. p. 4

[3] Rosenfeld A. Fuzzy Graphs, Fuzzy Sets and Their Applications. Cambridge, United States: Academic Press; 1975. pp. 77-95

[4] Atanassov K. Review and new results on intuitionistic fuzzy sets. International Journal Bioautomation. 2016;**20**:17-26

[5] Atanassov K. Intuitionistic Fuzzy Sets. Theory and Applications. New York: Physica-Verlag; 1999. DOI: 10.1007/978-3-7908-1870-3

[6] Atanassov KT. On Intuitionistic Fuzzy Sets Theory. 2012. p. 283. DOI: 10.1007/978-3-642-29127-2

[7] Mordeson J. Fuzzy line graphs. Pattern Recognition Letters. 1993;**14**(5): 381-384. DOI: 10.1016/0167-8655(93) 90115-T

[8] Firouzian S, Sedghi S, Shobe N. On edge fuzzy line graphs and their fuzzy congraphs. Control and Optimization in Applied Mathematics. 2021;**6**(1):81-89

[9] Akram M, Dudek W. Interval-valued fuzzy graphs. Computers and Mathematics with Applications. 2011;**61**:289-299. DOI: 10.1016/j.camwa.2010.11.004

[10] Naz S, Malik M, Rashmanlou H. Hypergraphs and transversals of hypergraphs in interval-valued intuitionistic fuzzy setting. Journal of Multiple-Valued Logic and Soft Computing. 2018;**30**:399-417

[11] Akram M. Interval-valued fuzzy line graphs. Neural Computing and Applications - NCA. 2012;**21**:1-6. DOI: 10.1007/s00521-011-0733-0

[12] Rashmanlou H, Borzooei R. New concepts of interval-valued intuitionistic (S, T)-fuzzy graphs. Journal of Intelligent and Fuzzy Systems. 2016; **30**(4):1893-1901

[13] Akram M, Davvaz B. Strong intuitionistic fuzzy graphs. Univerzitet u Nišu. 2012;**26**(1):177-196. DOI: 10.2298/ FIL1201177A

[14] Akram M, Parvathi R. Properties of intuitionistic fuzzy line graphs. Notes on Intuitionistic Fuzzy Sets. 2012;**18**(3): 52-60

[15] Parvathi R, Karunambigai MG, Atanassov KT. Operations on intuitionistic fuzzy graphs. In: 2009 IEEE International Conference on Fuzzy Systems. 20 August 2009. pp. 1396-1401. DOI:10.1109/FUZZY.2009.5277067

Chapter 6

Multi-Dimensional Codebooks for Multiple Access Schemes

Kais Hassan, Kosai Raoof and Pascal Chargé

Abstract

The sparse code multiple access (SCMA) scheme directly maps the incoming bits of several sources (users/streams) to complex multi-dimensional codewords selected from a specific predefined sparse codebook set. The codewords of all sources are then superimposed and exchanged. The shaping gain of the multi-dimensional constellation of SCMA leads to a better system performance. The decoder's objective will be to separate the superimposed sparse codewords. Most existing works on SCMA decoders employ message passing algorithm (MPA) or one of its variations, or a combination of MPA and other methods. The system architecture is highlighted and its basic principles are presented. Then, an overview of main multi-dimensional constellations for SCMA systems will be provided. Afterwards, we will focus on how the SCMA codebooks are decoded and how their performance is evaluated and compared.

Keywords: multi-dimensional constellations, codebook design, message passing algorithms, sparse code, code-domain

1. Introduction

The massive connectivity is one of the main requirements of the 5G telecommunication systems and beyond. One key to fulfill this objective is to allow several users to efficiently access the same resources (frequency band for example) simultaneously, this approach is called multiple access. Based on how the resources are shared among multiple users, two types of multiple access could be distinguished: orthogonal multiple access (OMA) and non-orthogonal multiple access (NOMA) [1].

A well-known OMA scheme is code-division multiple access (CDMA), the idea is to divide the symbol duration into a number of time slots or chips such that the spreading sequence associated to each user is chosen from a set of non sparse quasi-orthogonal ones. The overall transmitted sequence is the result of the superimposition of the symbols of all users which are spread over different chips. The number of served users is limited to the number of available quasi-orthogonal sequences, however, the orthogonality of sequences guarantees the simplicity of the receiver since a low complexity correlation operation is sufficient to detect the users' symbols despite the inter-sequence interference.

The key difference between code-domain NOMA techniques and the CDMA is that the spreading sequences of the former are restricted to non-orthogonal low cross-correlation sparse sequences such that more spreading sequences of code can be used, and consequently more users can be served simultaneously. One code-domain NOMA scheme which had shown to achieve a promising link level performance is SCMA [2]. Traditional code-domain schemes map bits to a symbol which is selected from one-dimensional constellation before spreading this symbol over a given low-density spreading sequence, SCMA combines together the two steps which gives birth to the idea of multi-dimensional constellations. The capacity to directly map the bits to some sparse SCMA codewords belonging to multi-dimensional codebooks attracts a lot of attention.

In this Chapter, we will present the SCMA system architecture by presenting its basic principles and its signal model. Then, existing methods for SCMA codebook design will be reviewed. Finally, we will explain how SCMA signal can be detected at the receiver either using the traditional MPA or one among its variations.

2. SCMA system architecture

In the following subsections, multi-dimensional coding principles are presented before illustrating why SCMA can be employed to provide multiple access.

2.1 Basic principles of multi-dimensional constellations

The SCMA spreads its sequence in the frequency domain over K subcarriers, these narrow frequency bands are also called resource elements (REs). For an uplink scenario, a base station (BS) serves simultaneously J separate users. The user j, so-called also layer j, sends a $K-$ dimensional codeword, $\mathbf{x}_j^{(m)}$, which represents $\log_2(M_j)$ data bits. Consequently, $\mathbf{x}_j^{(m)}$ must be chosen from a codebook, $\mathbf{C}_j$, of size M_j such that the multi-dimensional constellation, $\mathcal{C} = \{\mathbf{C}_j, 1 \leq j \leq J\}$ is designed to facilitate the multiple access. Actually, the codewords of all users, $\mathbf{x}_j^{(m)}, 1 \leq j \leq J$, are superimposed and exchanged over the K REs. In fact, $\mathcal{C}$ collects the signatures of served users. In order to increase the number of connected users, the codewords are designed to be sparse, i.e. all their entries are zeros except for few ones, in other words, the number of non-zero entries, N_j, must be lesser than the length of the codewords, K, i.e. $N_j \ll K$. Hence, the jth SCMA layer can be described by its *codebook sparsity degree*, N_j, and the whole SCMA system is characterized by

- d_f which is defined by the maximum degree of user superposition on a given RE,
- λ which denotes the overloading factor, λ is calculated by the ratio of number of users to number of REs, i.e. $\lambda = \frac{J}{K}$.

However, we must highlight that all the N_j non-zero entries of the codewords of $\mathbf{C}_j$ are located in the same positions.

Based on the different parameters of SCMA system, especially, the size of codebook of each user, M_j, and its codebook sparsity degree N_j, we can distinguish two kinds of SCMA system architectures:

i. The regular SCMA where users are treated equally, i.e. all users employ a codebook of size M and their signals are spread over N REs,

ii. The irregular SCMA, as its name indicates, is designed such that the codebooks are allocated differently according to the different needs of users [3].

Figure 1 presents an example of a regular SCMA system, obviously, we have $N_j = 2, 1 \leq j \leq J$ and $M_j = 4, 1 \leq j \leq J$. This system is characterized by $d_f = 3$ and $\lambda = 150\%$. In the rest of this chapter, a simple"SCMA" will refer implicitly to regular SCMA.

The received vector for an uplink SCMA system is given by,

$$\mathbf{y} = \sum_{j=1}^{J} \mathbf{H}_j \mathbf{x}_j^{(m)} + \mathbf{n}, \tag{1}$$

where $\mathbf{y} = (y_1, \cdots, y_K)^T$ and $\mathbf{x}_j^{(m)} = \left(x_{j,1}^{(m)}, \cdots, x_{j,K}^{(m)}\right)^T$. Let us denote the channel gain of user j on subcarrier k by $h_{j,k}$, hence the matrix $\mathbf{H}_j$ is diagonal of dimension $K \times K$ where $h_{j,k}, 1 \leq k \leq K$ are its diagonal entries. Finally, at the receiver, a zero-mean white circularly complex Gaussian noise, $\mathbf{n}$, with variance N_0 is added; i.e. $\mathbf{n} \sim \mathcal{CN}(0, N_0\mathbf{I}_K)$, where $\mathbf{I}_K$ is the identity matrix of size K.

3. SCMA codebook design

The design of SCMA codebook is usually based on several steps, a description of each one among them is given in this section. The idea is that the constellation function, associated with each user j generates a constellation set with M alphabets of length N. Then, the mapping matrix $\mathbf{V}_j$ maps the N-dimensional constellation points to SCMA codewords to form the codebook $\mathbf{C}_j$.

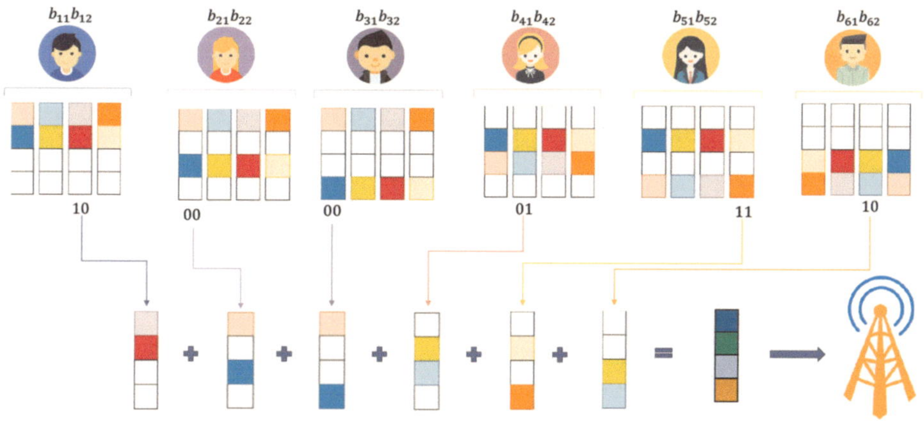

Figure 1.
The encoder of a regular SCMA system: The transmitted codeword is the superposition of the codeword of each user which is selected from its own codebook according to the $\log_2(M)$ *bit to be transmitted at each time frame.*

3.1 Codebook design procedure

The description of a SCMA system begins by determining the locations of non-zero elements of user $j, 1 \leq j \leq J$, via the vector $\mathbf{f}_{\mathrm{j}}$, for instance $\mathbf{f}_{\mathrm{j}} = [1,1,0,0]^T$ means that the user j employs the first two subcarriers only to send his data, this can be also described using another matrix, $\mathbf{V}_j$, of dimension $K \times N$, given by,

$$\mathbf{V}_j = \begin{bmatrix} 1 & 0 \\ 0 & 1 \\ 0 & 0 \\ 0 & 0 \end{bmatrix} \tag{2}$$

where $\mathbf{V}_j$ is the mapping matrix of user j. Thus, the whole SCMA system is described by gathering the $\mathbf{f}_{\mathrm{j}}$ vectors in one matrix $\mathbf{F}$ of dimension $K \times J$ such that $\mathbf{F} = (\mathbf{f}_1, \cdots, \mathbf{f}_{\mathrm{J}})$, $\mathbf{F}$ is called the factor graph matrix. The two matrices are related by $\mathbf{f}_j = \mathbf{V}_j \mathbf{V}_j^T$.

The factor graph matrix that represents the system in **Figure 1** is given by,

$$\mathbf{F} = \begin{bmatrix} 1 & 1 & 1 & 0 & 0 & 0 \\ 1 & 0 & 0 & 1 & 1 & 0 \\ 0 & 1 & 0 & 1 & 0 & 1 \\ 0 & 0 & 1 & 0 & 1 & 1 \end{bmatrix} \tag{3}$$

and the factor graph itself is depicted in **Figure 2** where every circle represents a user (so-called variable node) and every block represents a subcarrier (so-called function node).

Thus, the matrix $\mathbf{F}$ is related to the codeword $\mathbf{x}_j^{(m)}$, in Eq. (1), by the fact that the structure of $\mathbf{F}$ defines where zeros are located in the codebook from which the codeword $\mathbf{x}_j^{(m)}$ is selected.

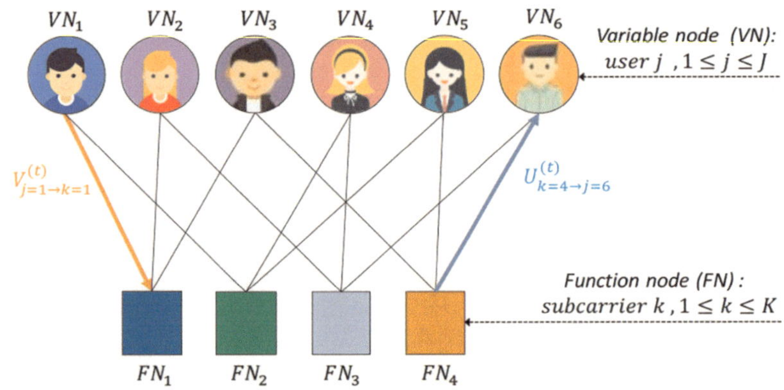

Figure 2.
The matrix, **F**, *can be translated into a factor graph. The encoder of the SCMA system illustrated in* **Figure 1** *is represented by this factor graph.*

As to the SCMA codebook design, it is considered as a joint optimization problem which objective is to find both the optimum user-to-RE mapping matrices $\mathcal{V}^*$ and the optimum multi-dimensional constellation $\mathcal{C}^*$, hence, this problem can be defined as,

$$\mathcal{V}^*, \mathcal{C}^* = \arg\max_{\mathcal{V},\mathcal{C}} D(\phi(\mathcal{V}, \mathcal{C}; J, M, N, K)) \tag{4}$$

where D is a design criterion and ϕ is the SCMA system as it was described above. However, the SCMA system must be designed under the assumption that J users are simultaneously connected, that is the system is fully loaded. In this case, the number of users is equal to the number of possible N-combinations among the K available REs, i.e. $J = \binom{K}{N}$. Hence, there is only one possible optimal mapping matrix solution.

Finding the optimum multi-dimensional constellation is still complex, one way to simplify this optimization problem is to divide it into several subproblems [4]. Hence, the multi-stage design of SCMA codebook is conducted in three main steps:

i. Firstly a constellation, $\mathbf{C}_{mc}$, composed of M words of size N is designed, $\mathbf{C}_{mc}$ is called the mother constellation,

ii. The mother constellation is considered as a seed from which user-specific multi-dimensional constellations are generated, this requires to design user-specific transformation matrices, $\mathbf{T}_j$,

iii. The combination of the above two steps gives a set of J matrices of size $M \times N$, the mapping matrix is employed to generate, $\mathcal{C}$, the set of J codebooks.

Taking into consideration the above-mentioned remarks, namely the uniqueness of the optimal solution for the mapping matrix and the multi-stage solution for the multi-dimensional constellation design, Eq. (3) can be rewritten as,

$$\left\{\mathbf{T}_j^*\right\}, \mathbf{C}_{mc}^* = \arg\max_{\{\mathbf{T}_j\}, \mathbf{C}_{mc}} D(\phi(\mathcal{V}^*, \{\mathbf{T}_j \mathbf{C}_{mc}\}; J, M, N, K)) \tag{5}$$

such that the jth codebook is calculated by,

$$\mathbf{C}_j = \mathbf{V}_j^* \mathbf{T}_j^* \mathbf{C}_{mc}^*. \tag{6}$$

In the following parts of this section, and inspired by the codebook design procedure illustrated in **Figure 3**, we present the major keys to design the mother constellation and the appropriate transformation operators.

3.2 Mother constellation design

The codebook of user J must be composed of M codewords since it encodes $\log_2(M)$ bits, each codeword has N non-zero elements. Hence, we start by designing a mother constellation matrix of N rows and M columns. Each row among the N ones represents a dimension among the N dimensions of the constellation. On the other hand, the mth column is a multi-dimensional point among the M multi-dimensional points of the constellation. The objective of the designing process is to guarantee a

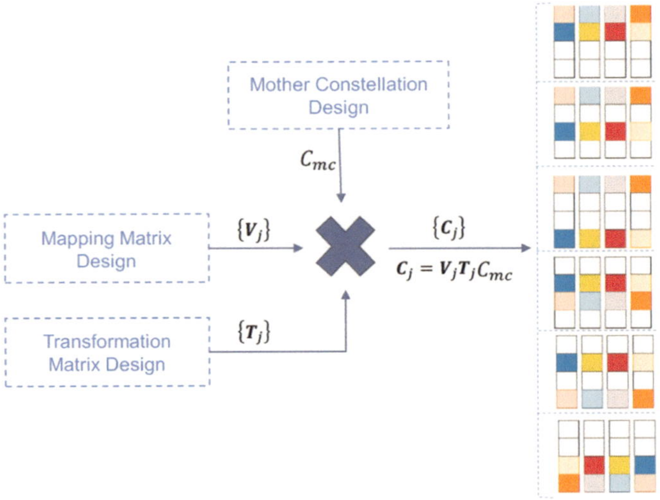

Figure 3.
A block diagram that illustrates the different steps which are conducted to design a SCMA codebook.

sufficiently good distance among all the points in the set C, that is keeping the points of the multi-dimensional constellation sufficiently far from each others such that they can be separated and decoded at the receiver. Consequently, the mother constellation must own a good distance profile. However, this requires to define how the distance between two multi-dimensional points is measured, this fundamentally defines the criterion D in eq. (4). Hereafter, the interested reader can find a list of the most employed distance definitions in the state of the art.

3.2.1 Euclidean distance

The Euclidean distance between two constellation points, $\mathbf{x}_i^{(u)}$ and $\mathbf{x}_j^{(m)}$, $1 \le u \le M$, $1 \le m \le M$, of user i and j respectively, $1 \le i \le J$, $1 \le j \le J$, is calculated by,

$$\mathrm{d}_E\left(\mathbf{x}_j^{(m)}, \mathbf{x}_i^{(u)}\right) = \|\mathbf{x}_j^{(m)} - \mathbf{x}_i^{(u)}\| \tag{7}$$

A classic design criterion is the minimum Euclidean distance of a multi-dimensional constellation [5, 6], it is defined as,

$$\mathrm{d}_E^{(\min)} = \min_{\substack{1 \le u, m \le M \\ 1 \le i, j \le J}} \left\{ d_E\left(\mathbf{x}_j^{(m)}, \mathbf{x}_i^{(u)}\right) \right\} \tag{8}$$

This criterion is more useful for evaluating the design of $\boldsymbol{C}_{mc}$ when all users are observing the same fading channel coefficients over their REs.

3.2.2 Euclidean kissing number

The key here is to count the number of distinct constellation point pairs which are separated by an Euclidean distance which is equal to the minimum Euclidean distance between any two points of the multi-dimensional constellation.

3.2.3 Product distance

The product distance between two N-dimensional complex constellation points, $\mathbf{x}_j^{(m)} = \left(x_{j,1}^{(m)}, \cdots, x_{j,N}^{(m)}\right)^T$ and $\mathbf{x}_i^{(u)} = \left(x_{i,1}^{(u)}, \cdots, x_{i,N}^{(u)}\right)^T$, is expressed as,

$$\mathrm{d}_P\left(\mathbf{x}_j^{(m)}, \mathbf{x}_i^{(u)}\right) = \prod_{\substack{1 \leq n \leq N \\ x_{j,n}^{(m)} \neq x_{i,n}^{(u)}}} |x_{j,n}^{(m)} - x_{i,n}^{(u)}| \tag{9}$$

The minimum product distance of a multi-dimensional constellation is given by,

$$\mathrm{d}_P^{(\min)} = \min_{\substack{1 \leq u, m \leq M \\ 1 \leq i, j \leq J}} \left\{ d_P\left(\mathbf{x}_j^{(m)}, \mathbf{x}_i^{(u)}\right) \right\} \tag{10}$$

This criterion is preferred when evaluating the design of C_{mc} in strong fading channel case, i.e., when channel coefficients over employed subcarriers are different.

3.2.4 Product kissing number

It is the number of distinct constellation point pairs with product distance equal to the minimum product distance.

To understand why SCMA performs well, the concept of shaping gain was introduced, the idea is to measure how the inherent shape of a multi-dimensional constellation, i.e. possessing additional dimensions in each constellation point or additional degrees of freedom, results in enhancing the distancing property of the SCMA constellation. We can assume that increasing the shaping gain means enhancing the overall system performance. The shaping gain is calculated by the ratio of the minimum distance between the points of multi-dimensional constellation to the minimum distance between the points of an one-dimensional one. The two constellations must have the same total power distributed on the same number of points. For instance, the authors in [5, 7] proposed 4-point two-dimensional mother constellation. The quadrature phase shift keying constellation is chosen as the reference one-dimensional constellation, the resulting shaping gain, which is calculated based on the Euclidean distance, is 1.25 dB.

Several new methods to design the SCMA mother constellation were proposed in the literature. In the following paragraphs, an overview of some interesting ones is presented, for each method the design criterion and the employed distance definition are highlighted.

The Euclidean distance will be intuitively the first to be used. One approach could be to fix a minimum Euclidean distance between any two points of the multi-dimensional constellation and to optimize another property, for instance the average constellation energy was minimized in [5], the resulting mother constellation is called the M-Beko. In [6], the authors proposed the M-Peng scheme which fixes the average energy and tries to maximize the minimum Euclidean distance between any two points of the alphabet.

Several research works aimed to reduce the number of superposing constellation points over each subcarrier or dimension as shown in **Figure 4**, however the users are

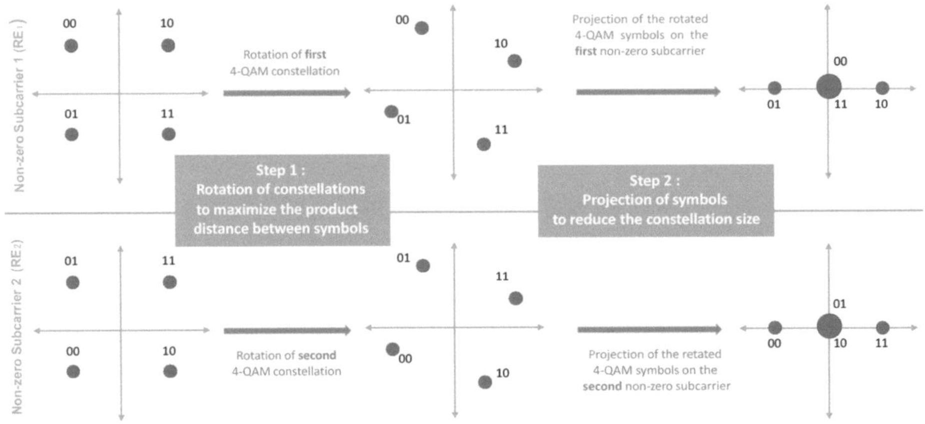

Figure 4.
Low-projection constellation: An example of QAM SCMA constellation points of size $M = 4$ with two non-zero REs, labeled based on gray coding. First step rotates the constellations to ensure a maximum product distance between symbols which enhances the detection process. The second step could better reduce the complexity of the receiver since some constellation points collide over each RE, for instance the constellation points corresponding to 00 *and* 11 *in the M-sized constellation collide over the first subcarrier, however, they have maximum distance over the second one which makes them separable using M_p-QAM constellation while $M_p \leq M$.*

distinct on other dimensions which will allow us to efficiently decode the codewords of each one among them [8–11]. This type of mother constellation is described as a low-projection one since it virtually reduces the codebook size from M into M_p where M_p is the size of the low-projected constellation. This leads to a further complexity reduction since the later is directly related to the *effective* codebook size, for instance, we can reduce the MPA complexity to $M_p^{d_f}$ instead of M^{d_f}. The low-projection approach is generally associated to the *product distance* criterion which has to be carefully adjusted to enhance the performance in the low signal-to-noise ratio (SNR) zone without compromising the performance in the high SNR one.

The design of constellations or code dictionaries is well studied in the state-of-the-art, we can mention, for example, digital modulation, CDMA, channel and source coding. This rich literature inspired some designs of multi-dimensional constellations. For instance, the authors in [9] proposed the T M QAM SCMA codebook whose design is based on the quadrature amplitude modulation (QAM). The idea is to design first two N-dimensional real constellations, then the N-dimensional complex constellation is conceived by applying a shuffling method on the Cartesian product of these N-dimensional real points. The optimization process is concluded by a rotation operation which aims at maximizing the minimum product distance of multi-dimensional constellation. The M LQAM scheme in [10] is a hybrid one between the shuffling method and the low-projection constellation approach. All the above-presented mother constellation designs did not take into consideration the wireless channel characteristics. For instance, the research work in [12, 13] derived a design criterion from cutoff rate of MIMO systems when the channel is assumed to be Rayleigh fading, the conceived constellation for SCMA systems is called M-Bao. In fact, the M multi-dimensional points of the T M QAM, M LQAM and M-Bao are based on the M corners of a $\log_2(M)$-dimensional hyper-cube. This inspired the authors in [14] to consider that the solution of the optimization problem in (4) is possible through an optimization of rotation angles of a hyper-cube, this method is denoted as M HQAM.

Analytical analysis showed that the complexity of MPA decoder is reduced from M^{d_f} to $(\log_2(M))^{d_f}$. Two examples of 2-dimensional mother constellations with 4-codewords are illustrated in **Table 1** and **Figure 5**, we hope that this will help the reader to understand the structure of a mother constellation.

Most of the above multi-dimensional constellations assume that the complex symbols can be randomly selected. Some propositions try to relax constraints on the research space by placing the constellation points of each dimension on multi-radius concentric rings [10, 15–17]. In [10], the symbols of each low-projection complex dimension are selected to form a M-point circular constellation, the M CQAM is based on the signal space diversity for MIMO systems over Rayleigh fading channels and results in a complexity reduction from M^{d_f} to $(M-1)^{d_f}$. The star-QAM constellation was proposed for digital modulation with the aim of being capable of flexibly adapting the ratios of multi-radius concentric rings. This approach was extended to multi-dimensional SCMA codebook design [15, 16]. The idea is to construct the first dimension of mother constellation from a star-QAM constellation of size M, afterwards, the following dimensions are deduced by applying some operations, for instance scaling and permuting, on the first dimension. The parameters of these operations are calculated through computer search which opens the door to designing constellations with large size and/or high dimension. An example of a constellation designed with this approach is represented in **Figure 6**. In [16], it was proposed to evaluate their proposition by directly applying the design criterion on the generated codewords of all users, contrary to other methods where it is only the mother constellation which was evaluated. The applied optimization criterion is the pairwise error probability between any two transmitted codewords $\mathbf{x}^{(1)}, \mathbf{x}^{(2)}$ which is given by,

$$\mathbb{P}\left(\mathbf{x}^{(1)}, \mathbf{x}^{(2)} | \mathbf{H}\right) = Q\left(\sqrt{\frac{\|\mathbf{H}(\mathbf{x}^{(1)} - \mathbf{x}^{(2)})\|^2}{2N_0}}\right). \tag{11}$$

where $\mathbf{H}$ is the uplink channel matrix of SCMA system as defined in Eq. (1).

In [17], each dimension of the mother constellation belongs to a ring such that the M complex points forms a uniformly spaced phase shift keying (PSK) constellation. Several dimensions mean several PSK rings with different radius values, hence the

Codeword m	**T4QAM**		**4LQAM**	
	$x_1^{(m)}$	$x_2^{(m)}$	$x_1^{(m)}$	$x_2^{(m)}$
1 (00)	$+\frac{3}{\sqrt{10}}$	$+\frac{1}{\sqrt{10}}$	$-\frac{\sqrt{2}}{2}$	$-\frac{\sqrt{2}}{2}i$
2 (01)	$-\frac{1}{\sqrt{10}}$	$+\frac{3}{\sqrt{10}}$	$-\frac{\sqrt{2}}{2}$	$+\frac{\sqrt{2}}{2}i$
3 (10)	$+\frac{1}{\sqrt{10}}$	$-\frac{3}{\sqrt{10}}$	$+\frac{\sqrt{2}}{2}$	$-\frac{\sqrt{2}}{2}i$
4 (11)	$-\frac{3}{\sqrt{10}}$	$-\frac{1}{\sqrt{10}}$	$+\frac{\sqrt{2}}{2}$	$+\frac{\sqrt{2}}{2}i$

Table 1.
This table presents T4QAM [9] and 4LQAM [10] mother constellations (4-codewords with 2 non-zeros dimensions) where $x_n^{(m)}$ belongs to dimension n, i.e. $x_n^{(m)}$ is the nth entry of the mth codeword m.

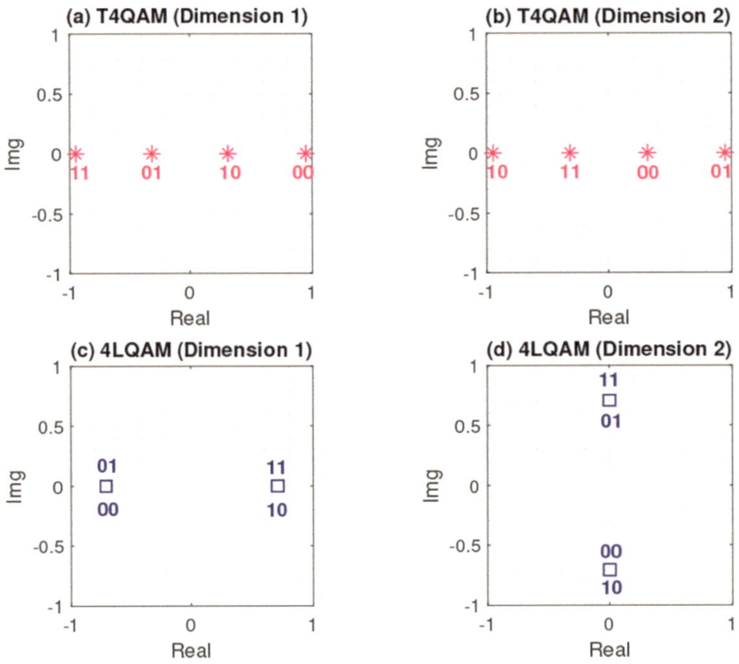

Figure 5.
Two examples of 2-dimensional mother constellations, namely T4QAM [9] and 4LQAM [10], each one is composed of 4-codewords: (a) the one-dimensional constellation of T4QAM as projected on the first dimension, (b) the one-dimensional constellation of T4QAM as projected on the second dimension, (c) the one-dimensional constellation of 4LQAM as projected on the first dimension and (d) the one-dimensional constellation of 4LQAM as projected on the second dimension.

multi-dimensional constellation is an amplitude and phase shift keying (APSK) constellation. In some applications, the PSK rings outperform square shaped QAM constellation since they provide a limited peak of power. The authors proposed a multi-stage optimization, the coded modulation capacity is employed as a design criterion for the first dimension of the mother constellation, then the other ones are optimized using permutations.

Once the mother constellation is designed, optimized and evaluated based on one of the above-presented design criteria, the applied transformations, which are used to generate the J codebooks, must be designed to preserve the characteristics of the mother constellation.

3.3 Transformation operators design

The design procedure of SCMA codebooks was introduced in **Figure 3**. First, the $N \times M$ mother constellation, $\mathbf{C}_{mc}$, is designed, then the sparse codebook of SCMA user j is constructed by applying a set of operators, $\mathbf{T}_j$, on $\mathbf{C}_{mc}$, and a mapping matrix, $\mathbf{V}_j$, as seen in eq. (5). Transforming a complex constellation can be conducted based on typical operations such as complex conjugate, rotation operator, interleaving and vector permutation. Several operators can be combined in some cases. Hence, the transformation operators must be chosen carefully such that the good characteristics of the mother constellation are conserved, their design was recently investigated [16–18, 20].

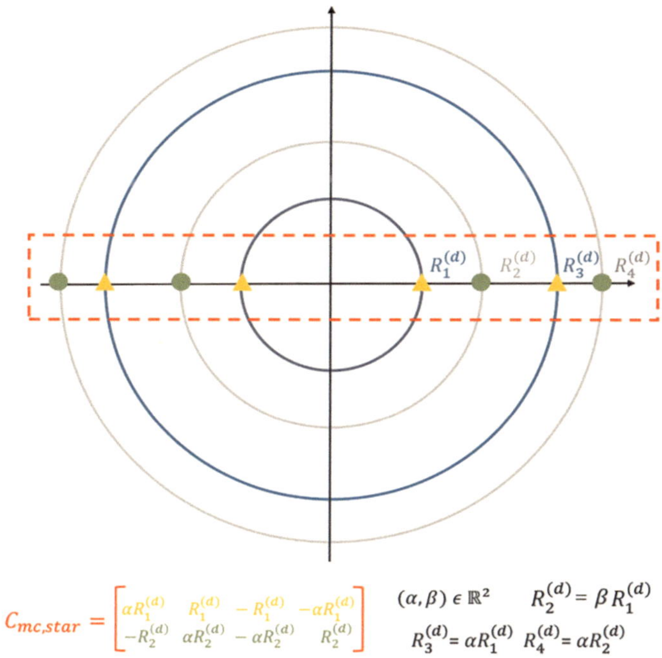

Figure 6.
A SCMA codebook of size $M = 4$ and sparsity degree $N = 2$ can be designed, for instance, based on a four-rings star-QAM mother constellation. Here, α and β are 2 reel design parameters.

The Euclidean distance is widely employed as a design criteria of the mother constellation. However, preserving its distancing property is possible by applying unitary rotation matrices [5–9, 19]. In this case, it is possible to merge the mapping operation and the transformation one to mold a new transformed factor graph matrix, $\mathbf{F}_T$. An example of a transformed factor graph matrix is expressed as,

$$\mathbf{F}_T = \begin{bmatrix} 0 & \varphi_1 & \varphi_2 & 0 & \varphi_3 & 0 \\ \varphi_2 & 0 & \varphi_3 & 0 & 0 & \varphi_1 \\ 0 & \varphi_2 & 0 & \varphi_1 & 0 & \varphi_3 \\ \varphi_1 & 0 & 0 & \varphi_3 & \varphi_2 & 0 \end{bmatrix} \tag{12}$$

where $\varphi_1 = e^{j\theta_1}, \varphi_2 = e^{j\theta_2}$ and $\varphi_3 = e^{j\theta_3}$. Traditionally, $\theta_1 = 0, \theta_2 = \frac{\pi}{3},$ and $\theta_3 = \frac{2\pi}{3}$. In this circumstance, the codebook of user 1, for instance, is calculated based on the following mapping and transformation matrices,

$$\mathbf{V}_1 = \begin{bmatrix} 0 & 0 \\ 1 & 0 \\ 0 & 0 \\ 0 & 1 \end{bmatrix} \quad \text{and} \quad \mathbf{T}_1 = \begin{bmatrix} e^{j\theta_2} & 0 \\ 0 & e^{j\theta_1} \end{bmatrix}.$$

It is worth noting that not only non-zero entries in each row of $\mathbf{F}_T$ are different but also those in each column, this property is called the Latin criterion. This means that the power variation and dimensional dependency can be controlled at the same time without compromising the Euclidean distance profile of the multi-dimensional mother constellation [8, 19]. **Figure 7** illustrates an example of SCMA system with 6 users ($J = 6$), their codebooks of size 4 ($M = 4$) are generated based on T4QAM mother constellation (which is described in **Table 1**) by employing unitary rotation matrices as described in eq. (11). This figure depicts how the constellation of each user is projected on each one between its associated two REs ($N = 2$). More details on these codebooks are given in Appendix A.

Multi-user codebooks generation can be further refined by optimizing computer-designed rotation matrices instead of the unitary rotation ones [12]. Furthermore, if the communication channel is assumed to be known, its phases can be exploited to extract random rotation angles which are used to generate the codebooks from the mother constellation. The authors in [18] combined the SCMA design with a form of codebook encryption which can ensure the link with low complexity. The rotation operations are not the only ones that can be employed. In [20], the transformation operator is designed based on a permutation set which is optimized to improve the detection reliability of the first decoded user, this largely improves the performance of SCMA receiver. The proposed design criterion tries to maximize the sum of distances among codewords which are multiplexed on the same RE (sum of distances per dimension). The factor graph matrix, **F**, defines the positions of non-zero elements of each user which are assumed to be fixed in the majority of SCMA designs as explained in subsection 3.1. One way to design transformation operator is to differentiate the non-zero locations according to the values of transmitted data bits which is considered as a permutation-based SCMA scheme [21]. This permutation approach does not suffer from a complexity overhead when compared to traditional one, however spectral efficiency does improve. This effort was extended by combining the matrix permutation and rotation operations to define the transformation operator as in [15, 16].

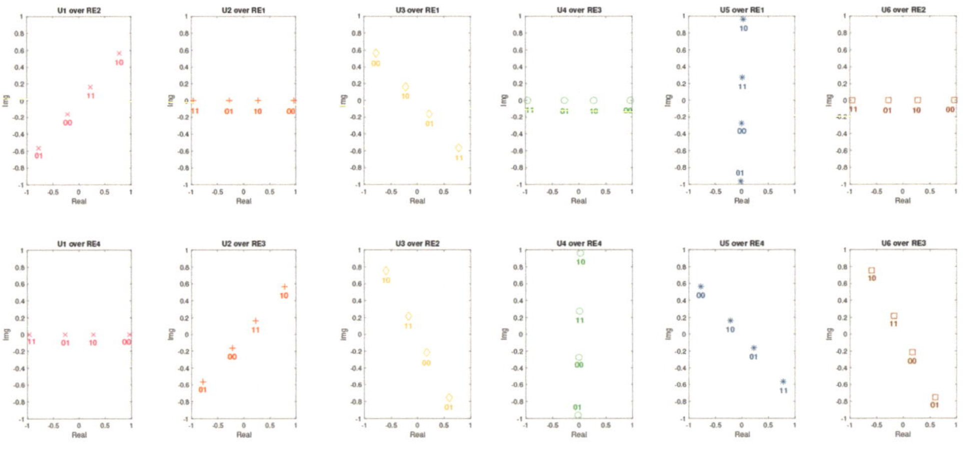

Figure 7.
An example of SCMA system with 6 users ($J = 6$), their codebooks of size 4 ($M = 4$) are generated based on T4QAM mother constellation by applying unitary rotation matrices, as described in eq. (11). This figure depicts how the constellation of each user is projected on each one between its associated two REs ($N = 2$).

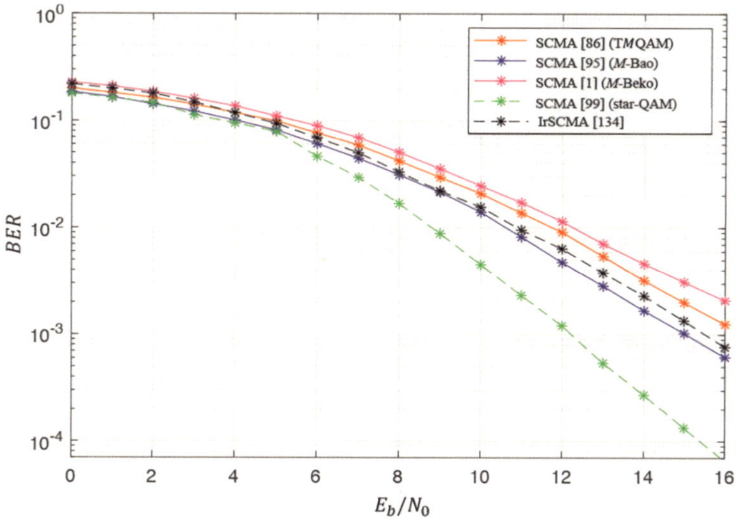

Figure 8.
BER as a function of SNR for SCMA system with different codebook designs: The number of orthogonal REs is 4, the number of users is 6, and the channel fading is assumed to be Rayleigh distributed.

Nevertheless, the objective is to minimize the rate of wrongly detected bits which is also called bit error rate (BER). For a given codeword error rate the BER depends on how codewords are labeled, hence it is obvious that choosing the appropriate labeling is an another aspect to be studied. In [17], the labeling was optimized to adjust the slope of the extrinsic information transfer (EXIT) chart.

The influence of the codebook design on the performance of a SCMA system is confirmed through simulations. **Figure 8** depicts the BER as a function of the SNR with different codebooks through Rayleigh fading channel. The star-QAM based SCMA design outperforms the other ones.

4. SCMA decoder design

It is worth mentioning that the SCMA encoder and decoder are two blocks among other ones at, respectively, the transmitter and receiver as explained in the block diagram illustrated in **Figure 9**. At the receiver side, the SCMA codewords must be segregated or decoded, this operation is preceded by the OFDM demodulation and the channel estimation, and followed by the deinterleaving and channel decoding.

At the transmitter, the bit to codeword mapping for each user is followed by the superimposition of the J mapped codewords which are selected from one of the above presented SCMA codebooks. At the receiver, the SCMA decoder aims to separate the

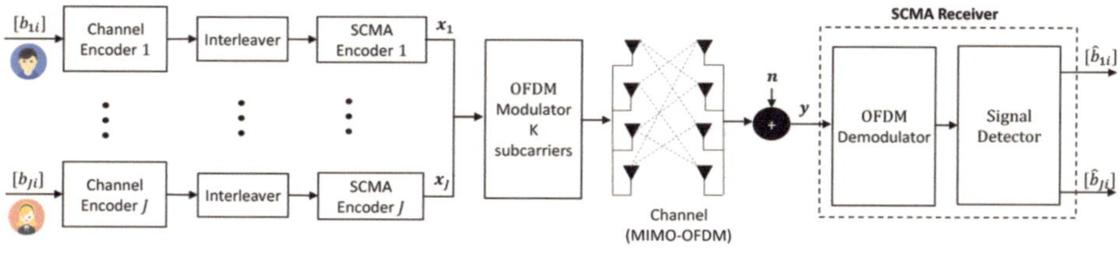

Figure 9.
This diagram illustrates the different essential blocks of the transmitter and receiver of SCMA system.

superimposed codewords despite that several users are occupying the same REs as described by the factor graph matrix. Most existing mechanisms employed for SCMA decoders are based on MPA or one of its variations, or its combination with other methods. In this section, we will present the basics of SCMA decoding.

4.1 Message passing algorithm

4.1.1 Traditional MPA

MPA is an iterative method based on passing some messages among concerned nodes. The nodes are the users, which are also considered as the variables nodes (VNs), and the subcarrier (or REs), which are also considered as the function nodes (FNs), and the massages are the extrinsic information among nodes, this mechanism is illustrated in **Figure 2**. Tha idea is that each FN calculates its outgoing message to a given VN depending on the incoming messages received from the reminder of VNs. The later ones will play the reciprocal role, that is each VN will reply by sending a message which is computed based on the received messages from the rest of FNs. This exchange among all the edges, i.e. all VNs and all FNs, is repeated at each iteration. After a given number of iterations, the bits of each user are estimated through the log-likelihood-rates (LLRs) of each coded bit. The MPA method is shown in Algorithm 1 and is based on three main steps: initialization, iterative message passing along edges and decision making.

Algorithm 1: Message Passing Algorithm.

Input: $\mathbf{y}, N_0, \mathbf{C}_j, h_j, j = 1, \cdots, J, N_{\text{iter}}$.
Estimation of the bits which were transmitted by each user. **Definitions.**
Users are represented by VNs, subcarriers are represented by FNs,

$$\mathcal{U}(k) = \{\text{all the VNs which are connected to } \text{FN}_k\}, k = 1, \cdots, K,$$
$$\mathcal{R}(j) = \{\text{all the FNs which are connected to } \text{VN}_j\}, j = 1, \cdots, J.$$

Step 1: Initialization.
Initially, each user expects to equally receive any codeword among the M ones:

$$V^0_{j\to k}\left(\mathbf{x}_j^{(m)}\right) = \mathbb{P}\left(\mathbf{x}_j^{(m)}\right) = \frac{1}{M}, j = 1, \cdots, J, k \in \mathcal{R}(j)$$

Step 2: Extrinsic information exchange among VNs and FNs

$$t \leq N_{\text{iter}}$$

1. The message to be sent from $\text{FN}_k, k = 1, \cdots, K$, to $\text{VN}_j, j \in \mathcal{U}(k)$, for each codeword $\mathbf{x}_j^{(m)} \in \mathbf{C}_j, m = 1, \cdots, M$, is computed by,

$$U^t_{k\to j}\left(\mathbf{x}_j^{(m)}\right) = \sum_{\mathbf{x}_i^{(m)} | i \in \mathcal{U}(k) \setminus j} \exp\left\{-\frac{1}{N_0}\|y_k - \sum_j h_{j,k} x_{j,k}^{(m)}\|^2\right\} \prod_{i \in \mathcal{U}(k) \setminus j} V^{t-1}_{i\to k}\left(\mathbf{x}_i^{(m)}\right)$$

2. The message to be sent from $\text{VN}_j, j = 1, \cdots, J$ to $\text{FN}_k, k \in \mathcal{R}(j)$, for each codeword $\mathbf{x}_j^{(m)} \in \mathbf{C}_j, m = 1, \cdots, M$, is calculated as,

$$V^t_{j\to k}\left(\mathbf{x}_j^{(m)}\right) = \frac{\prod_{i\in\mathcal{R}(j)\backslash k} U^{t-1}_{i\to j}\left(\mathbf{x}_j^{(m)}\right)}{\sum_{\mathbf{x}_j^{(l)}\in\mathbf{C}_j}\prod_{i\in\mathcal{R}(j)\backslash k} U^{t-1}_{i\to j}\left(\mathbf{x}_j^{(l)}\right)}.$$

It is essential to normalize this message in order to guarantee the numerical stability of MPA.

Step 3: Received bits estimation

1. The posteriori probability of each codeword for each user is represented by,

$$\mathbb{P}\left(\mathbf{x}_j^{(m)}\right) = \prod_{k\in\mathcal{R}(j)} U^{N_{\text{iter}}}_{k\to j}\left(\mathbf{x}_j^{(m)}\right), m = 1, \cdots, M, j = 1, \cdots, J.$$

2. Log-Likelihood-Rate for each coded bit, $b_i, 1 \le i \le \log_2(M)$, is given by,

$$\text{LLR}(b_i) = \log\left(\frac{\mathbb{P}(b_i = 0)}{\mathbb{P}(b_i = 1)}\right) = \log\left(\frac{\sum_{\left\{\mathbf{x}_j^{(m)}\in\mathbf{C}_j|b_i=0\right\}}\mathbb{P}\left(\mathbf{x}_j^{(m)}\right)}{\sum_{\left\{\mathbf{x}_j^{(m)}\in\mathbf{C}_j|b_i=1\right\}}\mathbb{P}\left(\mathbf{x}_j^{(m)}\right)}\right)$$

3. Finally, the value of each LLR is employed to decide on the corresponding bit as following,

$$\hat{b}_i = \begin{cases} 1 & \text{if } \text{LLR}(b_i) \le 0 \\ 0 & \text{otherwise.} \end{cases}$$

4.1.2 Variations of MPA

Despite being a referent decoder for SCMA, the complexity evaluation of MPA reveals that it relies on a large number of exponential calculus which are of high complexity. With the challenge to reduce this complexity and to fit with critical requirements of future wireless networks, several variations of MPA were proposed, among them we present here, the Max-Log-MPA and Log-MPA methods [22].

⊛ **Max-Log-MPA**: It is a simplified version of MPA based on a mathematical simplification which approximates the logarithm of a sum of exponential operations into a maximum operation. The key purpose is to move the iterative decoding process into logarithmic domain which eliminates the exponential terms in MPA by employing the simplified formula of *Jacobean logarithm*,

$$\log(\exp(a_1) + \ldots + \exp(a_n)) \approx \max(a_1, \ldots, a_n) \tag{13}$$

Thus, passing numerous messages from FNs to VNs, and vice versa, will be very less expensive in term of complexity. Based on (12), the expression of $\text{LLR}(b_i)$ presented in Algorithm 1 is modified as follows,

$$\mathrm{LLR}(b_i) = \max_{\left\{\mathbf{x}_j^{(m)} \in \mathbf{C}_j | b_i = 0\right\}} \left(\log\left(\mathbb{P}\left(\mathbf{x}_j^{(m)}\right)\right)\right) - \max_{\left\{\mathbf{x}_j^{(m)} \in \mathbf{C}_j | b_i = 1\right\}} \left(\log\left(\mathbb{P}\left(\mathbf{x}_j^{(m)}\right)\right)\right) \quad (14)$$

⊛ **Log-MPA**: The approximation of the *Jacobean logarithm formula* as presented in (12) makes the Max-Log-MPA a sub-optimal solution and results in a performance degradation. To mitigate this issue, a correction term was added by using another *Jacobean logarithm formula*. The adopted approximation is given by,

$$\log(\exp(a_1) + \ldots + \exp(a_n)) = a_j + \log\left(1 + \sum_{i \in \{1 \ldots n\} \backslash j} \exp(-|a_j - a_i|)\right) \quad (15)$$

where $a_j = \max(a_1, \ldots, a_n)$. Hence, the LLRs are further updated to be as below, rather than as in (13),

$$\begin{aligned}\mathrm{LLR}(b_i) = & \Bigg[\max_{\left\{\mathbf{x}_j^{(m)} \in \mathbf{C}_j | b_i = 0\right\}} \left(\log\left(\mathbb{P}\left(\mathbf{x}_j^{(m)}\right)\right)\right) + \\ & \log\left(1 + \sum_{m' \in \{1 \ldots M\} \backslash m} \exp\left(-\left|\log\left(\mathbb{P}\left(\mathbf{x}_j^{(m)}\right)\right) - \log\left(\mathbb{P}\left(\mathbf{x}_j^{(m')}\right)\right)\right|\right)\right)\Bigg] \\ & - \Bigg[\max_{\left\{\mathbf{x}_j^{(m)} \in \mathbf{C}_j | b_i = 1\right\}} \left(\log\left(\mathbb{P}\left(\mathbf{x}_j^{(m)}\right)\right)\right) + \\ & \log\left(1 + \sum_{m' \in \{1 \ldots M\} \backslash m} \exp\left(-\left|\log\left(\mathbb{P}\left(\mathbf{x}_j^{(m)}\right)\right) - \log\left(\mathbb{P}\left(\mathbf{x}_j^{(m')}\right)\right)\right|\right)\right)\Bigg]\end{aligned} \quad (16)$$

The performance of the above-presented variations of MPA, namely MPA, Log-MPA and Max-Log-MPA, were evaluated, **Figure 10** depicts the BER as a function of SNR through Rayleigh fading channel. The results show that the performance of Log-MPA is near-optimum when compared to that of MPA. The same is not valid for Max-Log-MPA, this can be explained by the correction term that was added to Log-MPA which obviously results in a performance compensation. On the other hand, the performance degradation of Log-MPA and Max-Log-MPA, due to the approximation used for each method, can be neglected in the high SNR zone. However, the Max-Log-MPA is still useful since it requires less computational effort when compared to Log-MPA which is still sufficiently complex to be considered challenging for energy-sensitive applications. It worth mentioning that reducing the computation complexity of the above-mentioned decoding methods is possible though reducing the value of d_f at the expense of largely constraining the codebook design.

Generally speaking, the complexity of MPA is intimately depending on the number of iterations which is usually one of its fixed parameters. This is not ideal since, on one hand, increasing the number of iterations will considerably increase the complexity, and on the other hand, not sufficiently iterating will lead to performance degradation. Hence, finding the near-optimal number of iterations is very useful. The performance of MPA as a function of SNR for different number of iterations is shown in **Figure 11**

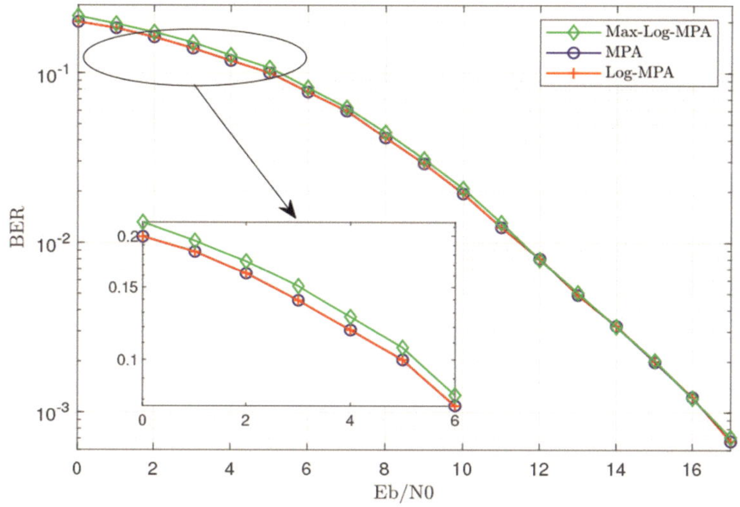

Figure 10.
BER as a function of SNR of MPA, log-MPA and MAX-log-MPA variations: The number of orthogonal REs is 4, the number of users is 6, and the channel fading is assumed to be Rayleigh distributed.

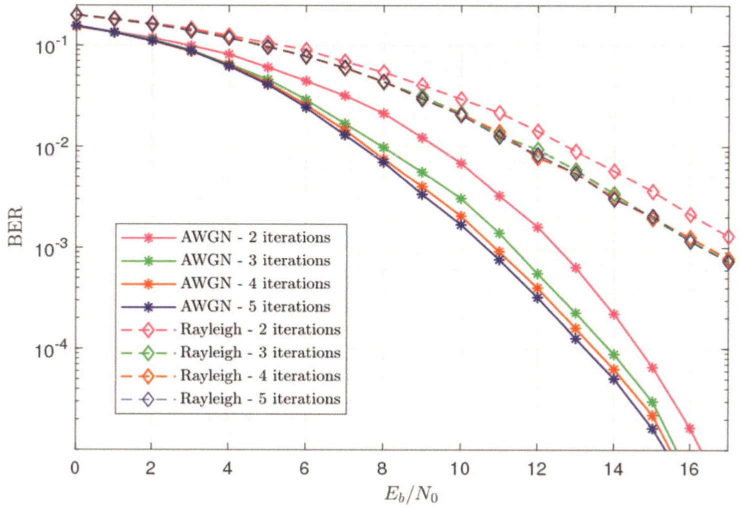

Figure 11.
Evaluation of the number of iterations on MPA performance: The number of orthogonal REs is 4, the number of users is 6, and the channel fading is assumed to be AWGN and Rayleigh distributed.

when the channel is assumed to be Gaussian or Rayleigh distributed. It is observed that the BER is lesser when the number of iterations increases under the two channel assumptions, nevertheless, beyond a certain limit, the performance improvement hits an upper bound. The same conclusions are valid for Log-MPA and Max-Log-MPA as reported in [22]. Therefore, a good compromise is to set the number of iterations to 4. Another approach is to supervise the convergence rate such that the number of iterations can be adjusted accordingly, the flexible number of iterations can be powerful when the convergence rate is efficiently measured.

5. Conclusions

In this Chapter, we presented the structure and basic principles of SCMA. Then, SCMA encoder and decoder designs were reviewed through their most known techniques. A simulations-based comparison among different existing approaches for codebook design as well as for signal decoding was conducted.

Nomenclature

BER	Bit error rate
BS	Base station
CDMA	Code-division multiple access
EXIT	Extrinsic information transfer
FN	Function node
LLR	Log-likelihood-rate
MPA	Message passing algorithm
NOMA	Non-orthogonal multiple access
OMA	Orthogonal multiple access
PSK	Phase shift keying
QAM	Quadrature amplitude modulation
RE	Resource elements
SCMA	Sparse code multiple access
SNR	Signal-to-noise ratio
VN	Variables node

A. Appendix

To better illustrates SCMA mapping, a numerical example of complete SCMA codebooks, as depicted in **Figures 5** and 7, is provided in the following (**Figure 12**).

$$\mathbf{CB}_1 = \left(\begin{bmatrix} 0 \\ -0.182 - 0.132i \\ 0 \\ +0.785 \end{bmatrix} \begin{bmatrix} 0 \\ -0.635 - 0.462i \\ 0 \\ -0.224 \end{bmatrix} \begin{bmatrix} 0 \\ +0.635 + 0.462i \\ 0 \\ +0.224 \end{bmatrix} \begin{bmatrix} 0 \\ +0.182 + 0.132i \\ 0 \\ -0.785 \end{bmatrix} \right)$$

$$\mathbf{CB}_2 = \left(\begin{bmatrix} +0.785 \\ 0 \\ -0.182 - 0.132i \\ 0 \end{bmatrix} \begin{bmatrix} -0.224 \\ 0 \\ -0.635 - 0.462i \\ 0 \end{bmatrix} \begin{bmatrix} +0.224 \\ 0 \\ +0.635 + 0.462i \\ 0 \end{bmatrix} \begin{bmatrix} -0.785 \\ 0 \\ +0.182 + 0.132i \\ 0 \end{bmatrix} \right)$$

$$\mathbf{CB}_3 = \left(\begin{bmatrix} -0.635 + 0.462i \\ +0.139 - 0.176i \\ 0 \\ 0 \end{bmatrix} \begin{bmatrix} +0.182 - 0.132i \\ +0.487 - 0.616i \\ 0 \\ 0 \end{bmatrix} \begin{bmatrix} -0.182 + 0.132i \\ -0.487 + 0.616i \\ 0 \\ 0 \end{bmatrix} \begin{bmatrix} +0.635 - 0.462i \\ -0.139 + 0.176i \\ 0 \\ 0 \end{bmatrix} \right)$$

$$\mathbf{CB}_4 = \left(\begin{bmatrix} 0 \\ 0 \\ +0.785 \\ -0.006 - 0.224i \end{bmatrix} \begin{bmatrix} 0 \\ 0 \\ -0.224 \\ -0.019 - 0.785i \end{bmatrix} \begin{bmatrix} 0 \\ 0 \\ +0.224 \\ +0.019 + 0.785i \end{bmatrix} \begin{bmatrix} 0 \\ 0 \\ -0.785 \\ +0.006 + 0.224i \end{bmatrix} \right)$$

$$\mathbf{CB}_5 = \left(\begin{bmatrix} -0.006 - 0.224i \\ 0 \\ 0 \\ -0.635 + 0.462i \end{bmatrix} \begin{bmatrix} -0.019 - 0.785i \\ 0 \\ 0 \\ +0.182 - 0.132i \end{bmatrix} \begin{bmatrix} +0.019 + 0.785i \\ 0 \\ 0 \\ -0.182 + 0.132i \end{bmatrix} \begin{bmatrix} +0.006 + 0.224i \\ 0 \\ 0 \\ +0.635 - 0.462i \end{bmatrix} \right)$$

$$\mathbf{CB}_6 = \left(\begin{bmatrix} 0 \\ +0.785 \\ +0.139 - 0.176i \\ 0 \end{bmatrix} \begin{bmatrix} 0 \\ -0.224 \\ +0.487 - 0.616i \\ 0 \end{bmatrix} \begin{bmatrix} 0 \\ +0.224 \\ -0.487 + 0.616i \\ 0 \end{bmatrix} \begin{bmatrix} 0 \\ -0.785 \\ -0.139 + 0.176i \\ 0 \end{bmatrix} \right)$$

Figure 12.
The 2-dimensional codebooks with 4-codewords, generated for $J = 6$ users, as described in ***Figure 7***.

Author details

Kais Hassan[1*], Kosai Raoof[1] and Pascal Chargé[2]

1 Laboratoire d'Acoustique de l'Université du Mans (LAUM), Le Mans University, Le Mans, France

2 Institut d'Electronique et des Technologies du numéRique (IETR), Polytech Nantes, Graduate School of Engineering of Nantes University, Nantes, France

*Address all correspondence to: kais.hassan@univ-lemans.fr

References

[1] Rebhi M, Hassan K, Raoof K, Chargé P. Sparse code multiple access: Potentials and challenges. IEEE Open Journal of the Communications Society. 2021;**2**:1205-1238

[2] Rebhi M, Hassan K, Raoof K, Chargé P. Deep learning for a fair distance-based SCMA detector. In: Proceedings of IEEE Wireless Communications and Networking Conference (WCNC). Austin-Texas, USA; April 2022

[3] Rebhi M, Hassan K, Raoof K, Chargé P. An adaptive uplink SCMA scheme based on channel state information. In: Proceedings of URSI, Future Network: 5G beyond Workshop. Paris, France; Mars 2020. pp. 1-7

[4] Vameghestahbanati M, Marsland ID, Gohary RH, Yanikomeroglu H. Multidimensional constellations for uplink scma systems—A comparative study. IEEE Communications Surveys Tutorials. 2019;**21**(3):2169-2194. DOI: 10.1109/COMST.2019.2910569

[5] Beko M, Dinis R. Designing good multi-dimensional constellations. IEEE Wireless Communications Letters. 2012; **1**(3):221-224

[6] Peng J, Chen W, Bai B, Guo X, Sun C. Joint optimization of constellation with mapping matrix for SCMA codebook design. IEEE Signal Processing Letters. 2017;**24**(3):264-268. DOI: 10.1109/LSP.2017.2653845

[7] Nikopour H, Baligh H. Sparse code multiple access. In: Proceedings of IEEE 24th Annual International Symposium on Personal, Indoor, and Mobile Radio Communications (PIMRC), London, UK. 2013. pp. 332-336

[8] Taherzadeh M, Nikopour H, Bayesteh A, Baligh H. Scma codebook design. In: Proceedings of 2014 IEEE 80th Vehicular Technology Conference (VTC2014-Fall), Vancouver, BC, Canada. 2014. pp. 1-5

[9] Taherzadeh M, Nikopour H, Bayesteh A, Baligh A. System. Method for Designing and Using Multidimensional Constellations. USA: U.S Patent 9,509,379; November 2016

[10] Metkarunchit T. Scma codebook design base on circular-qam. In: Proceedings of 2017 Integrated Communications, Navigation and Surveillance Conference (ICNS), Herndon, VA, USA. 2017. pp. 3E1–1-3E1–8

[11] Wei F, Chen W. Low complexity iterative receiver design for sparse code multiple access. IEEE Transactions on Communications. 2017;**65**(2):621-634

[12] Bao J, Ma Z, Ding Z, Karagiannidis GK, Zhu Z. On the design of multiuser codebooks for uplink sCMA systems. IEEE Communications Letters. 2016;**20**(10):1920-1923

[13] Bao J, Ma Z, Mahamadu MA, Zhu Z, Chen D. Spherical codes for scma codebook. In: Proceedings of 2016 IEEE 83rd Vehicular Technology Conference (VTC Spring), Nanjing, China. 2016. pp. 1-5

[14] Vameghestahbanati M. Hypercube-based multidimensional constellation design for uplink scma systems. In: Proceedings of 2020 IEEE International Conference on Communications (ICC), Dublin, Ireland. 2020

[15] Yu L, Lei X, Fan P, Chen D. An optimized design of scma codebook

based on star-qam signaling constellations. In: Proceedings of 2015 International Conference on Wireless Communications Signal Processing (WCSP), Nanjing, China. 2015. pp. 1-5

[16] Yu L, Fan P, Cai D, Ma Z. Design and analysis of scma codebook based on star-qam signaling constellations. IEEE Transactions on Vehicular Technology. 2018;**67**(11):10543-10553. DOI: 10.1109/TVT 2018.2865920

[17] Bao J, Ma Z, Xiao M, Tsiftsis TA, Zhu Z. Bit-interleaved coded scma with iterative multiuser detection: Multidimensional constellations design. IEEE Transactions on Communications. 2018;**66**(11):5292-5304

[18] Lai K, Lei J, Wen L, Chen G, Li W, Xiao P. Secure transmission with randomized constellation rotation for downlink sparse code multiple access system. IEEE Access. 2018;**6**:5049-5063

[19] Xiao K, Xia B, Chen Z, Xiao B, Chen D, Ma S. On capacity-based codebook design and advanced decoding for sparse code multiple access systems. IEEE Transactions on Wireless Communications. 2018;**17**(6):3834-3849

[20] Yan C, Kang G, Zhang N. A dimension distance-based scma codebook design. IEEE Access. 2017;**5**: 5471-5479

[21] Kulhandjian M, D'Amours C. Design of permutation-based sparse code multiple access system. In: Proceedings of 2017 IEEE 28th Annual International Symposium on Personal, Indoor, and Mobile Radio Communications (PIMRC), Montreal, QC, Canada. 2017. pp. 1-6

[22] Ameur WB, Mary P, Dumay M, Hélard J, Schwoerer J. Performance study of mpa, log-mpa and max-log-mpa for an uplink scma scenario. In: Proceedings of 26th International Conference on Telecommunications (ICT), Hanoi, Vietnam. 2019. pp. 411-416

Chapter 7

Polynomials Related to Generalized Fibonacci Sequence

Manjeet Singh Teeth and Sanjay Harne

Abstract

The Fibonacci polynomials are a polynomial sequence that can be considered as a generalization of the Fibonacci numbers. Fibonacci polynomials are defined by a recurrence relation: $F_n(x) = xF_{n-1}(x) + F_{n-2}(x), n \geq 2$ where $F_0 = 0, F_1 = 1$. The first few Fibonacci polynomials are $F_0 = 0$, $F_0(x) = 0$, $F_1(x) = 1$, $F_2(x) = x$, $F_3(x) = x^2 + 1$. In this chapter, we extend the Fibonacci recurrence relation to define the sequence $\{K_n\}$ and will derive some properties of this sequence. We also define four comparison sequences $\{P_n\}$, $\{Q_n\}$, $\{R_n\}$, and $\{S_n\}$ and obtain some identities with the help of generating matrix.

Keywords: Fibonacci numbers, Fibonacci sequence, generating matrix, rabbit problem, Polynomials

1. Introduction

The Fibonacci sequence [1] receives its name from Leonardo Pisano, known as Fibonacci who was the most talented Italian mathematician of middle age. It is supposed that he was the first mathematician who introduced the Hindu-Arabic system of numbers to Italians. His work 'Liber-Abaci' (1202) is famous for this.

In the Liber Abaci, Leonardo states the famous "Rabbit Problem" for attaining the output of this rabbit problem.

1.1 Utilization of Fibonacci sequence in the study of famous rabbit problem

"How many pairs of rabbits are born of one pair in a year?" This problem is stated in the form: "Suppose a newly-born pair of rabbits, one male and one female, are put in a field. Rabbits are able to mate at the age of 1 month so that at the end of its second month a female can produce another pair of rabbits."

Suppose that our rabbits never die and that the female always produces one new pair (one male and one female) every month from the second month on.

Leonardo also gave the solution to this problem and obtained the sequence of numbers as a result:

$$1, 1, 2, 3, 5, 8, \ldots$$

This sequence is called the Fibonacci sequence. The Fibonacci sequence is defined by the recurrence relation as,

$$F_n = F_{n-1} + F_{n-2}, n > 1$$

Waddilli, M.E. [2] has extended the Fibonacci recurrence relation to define the sequence $\{K_n\}$, where,

$$K_n = K_{n-1} + K_{n-2} + K_{n-3}, n > 3 \quad (1)$$

where, K_0, K_1, K_2 are given arbitrary algebraic integers.

Jaiswal, D.V. [3] has extended Fibonacci recurrence relation to define the sequence $\{Q_0\}$, where,

$$Q_n = Q_{n-1} + Q_{n-2} + Q_{n-3} + Q_{n-4}, n > 4 \quad (2)$$

where, Q_0, Q_1, Q_2 are given arbitrary algebraic integers.

Harne, S. [4] has extended Fibonacci recurrence relation to define the sequence $\{D_n\}$, where,

$$D_n = D_{n-1} + D_{n-2} + D_{n-3} + D_{n-4} + Q_{n-5}, n > 5 \quad (3)$$

where, D_0, D_1, D_2 are given arbitrary algebraic integers.

In this chapter, Teeth MS. [5] shall further extend the Fibonacci recurrence relation [6–10] to define the sequence $\{C_n\}$ and shall discuss some properties of this sequence. We shall also consider the four comparison sequences $\{P_n\}$, $\{Q_n\}$, $\{R_n\}$, and $\{K_n\}$.

2. The generalized sequence as per our propose model $\{K_n\}$

We consider the following sequence,

$$\{C_n\} = C_0, C_1, C_2, C_3, \ldots., C_n$$

where, $C_0, C_1, C_2, C_3, C_4, C_5, C_0$ are arbitrary algebraic integers all of which are not zero and

$$C_n = C_{n-1} + C_{n-2} + C_{n-3} + C_{n-4} + C_{n-5} + C_{n-6}, n \geq 6 \quad (4)$$

We also consider the sequence $\{P_n\} = P_0, P_1, P_2, P_3, \ldots., P_n.$
where,

$$\begin{aligned} P_0 &= C_3 - C_2 - C_1 - C_0 \\ P_1 &= C_4 - C_3 - C_2 - C_1 \\ P_2 &= C_5 - C_4 - C_3 - C_2 \\ P_3 &= C_6 - C_5 - C_4 - C_3 \\ P_4 &= C_7 - C_6 - C_5 - C_4 \end{aligned} \quad (5)$$

$$\text{with, } P_n = C_{n-1} + C_{n-2} + C_{n-3} + C_{n-4} + C_{n-5}, n \geq 5 \quad (6)$$

and $\{Q_n\} = Q_0, Q_1, Q_2, Q_3, \ldots \ldots., Q_n$
where,

$$
\begin{aligned}
Q_0 &= C_4 - C_3 - C_2 - C_1 \\
Q_1 &= C_5 - C_4 - C_3 - C_2 - C_1 \\
Q_2 &= C_6 - C_5 - C_4 - C_3 - C_2
\end{aligned}
\tag{7}
$$

$$\text{with,} \qquad Q_n = C_{n-1} + C_{n-2} + C_{n-3} + C_{n-4} \tag{8}$$

and $\{R_n\} = R_0, R_1, R_2, R_3, \ldots ., R_n$ where,

$$
\begin{aligned}
R_0 &= C_5 - C_4 - C_3 - C_2 - C_1 - C_0 \\
R_1 &= C_6 - C_5 - C_4 - C_3 - C_2 - C_1 \\
R_2 &= C_7 - C_6 - C_5 - C_4 - C_3 - C_2 \\
R_3 &= C_8 - C_7 - C_6 - C_5 - C_4 - C_3 \\
R_4 &= C_9 - C_8 - C_7 - C_6 - C_5 - C_4
\end{aligned}
\tag{9}
$$

$$\text{with,} R_n = C_{n-1} + C_{n-2} + C_{n-3} \tag{10}$$

and $\{S_n\} = S_0, S_1, S_2, S_3, \ldots ., S_n$ where,

$$
\begin{aligned}
S_0 &= C_6 - C_5 - C_4 - C_3 - C_2 - C_1 - C_0 \\
S_1 &= C_7 - C_6 - C_5 - C_4 - C_3 - C_2 - C_1 \\
S_2 &= C_8 - C_7 - C_6 - C_5 - C_4 - C_3 - C_2 \\
S_3 &= C_9 - C_8 - C_7 - C_6 - C_5 - C_4 - C_3 \\
S_4 &= C_{10} - C_9 - C_8 - C_7 - C_6 - C_5 - C_4
\end{aligned}
\tag{11}
$$

$$\text{with,} S_n = C_{n-1} + C_{n-2}, n \geq 2 \tag{12}$$

From (4) and (6) we have for $n \geq 11$

$$
\begin{aligned}
P_n = \ & C_{n-2} + C_{n-3} + C_{n-4} + C_{n-5} + C_{n-6} + C_{n-7} + C_{n-3} + C_{n-4} C_{n-5} + C_{n-6} \\
& + C_{n-7} + C_{n-8} + C_{n-4} + C_{n-5} + C_{n-6} + C_{n-7} + C_{n-8} + C_{n-9} + C_{n-5} + C_{n-6} \\
& + C_{n-7} + C_{n-8} + C_{n-9} + C_{n-10} + C_{n-6} + C_{n-7} + C_{n-8} + C_{n-9} + C_{n-10} + C_{n-11} \\
P_n &= P_{n-1} + P_{n-2} + P_{n-3} + P_{n-4} + P_{n-5} + P_{n-6}
\end{aligned}
$$

Now, from Eqs. (5) and (6),

$$
\begin{aligned}
P_{10} = \ & (C_8 + C_7 + C_6 + C_5 + C_4) + (C_7 + C_6 + C_5 + C_4 + C_3) \\
& + (C_6 + C_5 + C_4 + C_3 + C_2) + (C_5 + C_4 + C_3 + C_2 + C_1) \\
& + (C_4 + C_3 + C_2 + C_1 + C_0) + (C_7 - C_6 - C_5 - C_4) \\
P_{10} &= P_9 + P_8 + P_7 + P_6 + P_5 + P_4 \\
\text{Similarly,} P_9 &= P_8 + P_7 + P_6 + P_5 + P_4 + P_3 \\
P_8 &= P_7 + P_6 + P_5 + P_4 + P_3 + P_2 \\
P_7 &= P_6 + P_5 + P_4 + P_3 + P_2 + P_1
\end{aligned}
$$

Hence, we have for $n \geq 6$

$$P_n = P_{n-1} + P_{n-2} + P_{n-3} + P_{n-4} + P_{n-5} + P_{n-6} \tag{13}$$

Proceeding on similar lines, it can be shown that for $n \geq 6$.

$$\begin{aligned} Q_n &= C_{n-2} + C_{n-3} + C_{n-4} + C_{n-5} + C_{n-6} + C_{n-7} \\ &+C_{n-3} + C_{n-4} + C_{n-5} + C_{n-6} + C_{n-7} + C_{n-8} \\ &+C_{n-4} + C_{n-5} + C_{n-6} + C_{n-7} + C_{n-8} + C_{n-9} \\ &+C_{n-5} + C_{n-6} + C_{n-7} + C_{n-8} + C_{n-9} + C_{n-10} \\ Q_n &= Q_{n-1} + Q_{n-2} + Q_{n-3} + Q_{n-4} + Q_{n-5} + Q_{n-6}, n \geq 6 \end{aligned} \tag{14}$$

Proceeding on similar lines it can be shown that for $n \geq 6$

$$\begin{aligned} R_n &= C_{n-2} + C_{n-3} + C_{n-4} + C_{n-5} + C_{n-6} + C_{n-7} \\ &+C_{n-3} + C_{n-4} + C_{n-5} + C_{n-6} + C_{n-7} + C_{n-8} \\ &+C_{n-4} + C_{n-5} + C_{n-6} + C_{n-7} + C_{n-8} + C_{n-9} \\ R_n &= R_{n-1} + R_{n-2} + R_{n-3} + R_{n-4} + R_{n-5} + R_{n-6}, n \geq 6 \end{aligned} \tag{15}$$

Proceeding on similar lines it can be shown that for $n \geq 6$

$$\begin{aligned} S_n &= C_{n-2} + C_{n-3} + C_{n-4} + C_{n-5} + C_{n-6} + C_{n-7} \\ &+C_{n-3} + C_{n-4} + C_{n-5} + C_{n-6} + C_{n-7} + C_{n-8}, n \geq 6 \\ S_n &= S_{n-1} + S_{n-2} + S_{n-3} + S_{n-4} + S_{n-5} + S_{n-6}, n \geq 6 \end{aligned} \tag{16}$$

Thus, the four sequences $\{P_n\}$, $\{Q_n\}$, $\{R_n\}$, and $\{S_n\}$ are special cases of sequence $\{C_n\}$ and all obtained by taking different initial values [11, 12].

On taking,

$$\begin{gathered} C_0 = C_1 = C_2 = 0, C_3 = C_4 = 1, C_5 = 2 \\ C_0 = C_1 = 0, C_2 = 1, C_3 = 0, C_4 = 1, C_5 = 2 \\ C_0 = 0, C_1 = 1, C_2 = C_3 = 0, C_4 = 1, C_5 = 2 \\ C_0 = 1, C_1 = C_2 = C_3 = 0, C_4 = 1, C_5 = 2 \\ C_0 = C_1 = C_2 = C_3 = 0, C_4 = 1, C_5 = 2 \\ 0,0,0,1,1,2,4,8,16,32,63, \ldots J_n, \ldots \\ 0,0,1,0,1,2,4,8,16,31,62, \ldots K_n, \ldots \\ 0,1,0,0,1,2,4,8,15,30,59, \ldots L_n, \ldots \\ 1,0,0,0,1,2,4,7,14,28,56, \ldots M_n, \ldots \\ 0,0,0,0,1,2,3,6,12,24,48, \ldots N_n, \ldots \end{gathered} \tag{17}$$

Here, we find that

$$\begin{aligned} K_n &= J_{n-1} + J_{n-2} + J_{n-3} + J_{n-4} + J_{n-5} \\ L_n &= J_{n-1} + J_{n-2} + J_{n-3} + J_{n-4} \\ M_n &= J_{n-1} + J_{n-2} + J_{n-3} \\ N_n &= J_{n-1} + J_{n-2} \end{aligned}$$

Hence, we say that $\{J_n\}$ is C_n type sequence, while $\{K_n\}$ is P_n type sequence, and $\{L_n\}$ is Q_n type sequence, while $\{M_n\}$ is R_n type sequence, and $\{N_n\}$ is S_n type sequence.

2.1 Linear sums and some properties

We have derived simple properties [2, 13, 14] of the sequences $\{C_n\}$, $\{P_n\}$, $\{Q_n\}$, $\{R_n\}$, and$\{S_n\}$, expressing each of the terms $C_6, C_7, C_8, \ldots, C_{n+5}C_6$, as the sum of its six preceding terms, as given in (4) adding both sides we obtained on
Simplification:

$$\sum_{i=0}^{n} C_i = \frac{1}{5}\{C_{n+5} - C_{n+3} - 2C_{n+2} - 3C_{n+1} + C_n - (C_5 - C_3 - 2C_2 - 3C_1 - 4C_0)\} \tag{18}$$

On using (4), (5), (7), (9), and (12), we get

$$\sum_{i=0}^{n} C_{6i} = \sum_{i=0}^{6n-1} C_i + C_0 \tag{19}$$

$$\sum_{i=0}^{n} C_{6i+2} = \sum_{i=0}^{6n+1} C_i + P_0 \tag{20}$$

$$\sum_{i=0}^{n} C_{6i+3} = \sum_{i=0}^{6n+2} C_i + Q_0 \tag{21}$$

$$\sum_{i=0}^{n} C_{6i+4} = \sum_{i=0}^{6n+3} C_i + R_0 \tag{22}$$

$$\sum_{i=0}^{n} C_{6i+5} = \sum_{i=0}^{6n+4} C_i + S_0 \tag{23}$$

$$\sum_{i=0}^{n} C_{6i+6} = \sum_{i=0}^{6n+5} C_i + (S_1 - C_0) \tag{24}$$

$$\sum_{i=0}^{n} C_{6i+5} = \sum_{i=0}^{6n+4} C_i + (R_1 - C_0) \tag{25}$$

$$\sum_{i=0}^{n} C_{6i+4} = \sum_{i=0}^{6n+3} C_i + (Q_1 - C_0) \tag{26}$$

$$\sum_{i=0}^{n} C_{6i+3} = \sum_{i=0}^{6n+2} C_i(P_1 - C_0) \tag{27}$$

2.2 Property of sequence $\{J_{n-2}\}$

Theorem: For the sequence $\{J_n\}$ we have,

$$\begin{vmatrix} J_n & J_{n+1} & J_{n+2} & J_{n+3} & J_{n+4} & J_{n+5} \\ J_{n+1} & J_{n+2} & J_{n+3} & J_{n+4} & J_{n+5} & J_{n+6} \\ J_{n+2} & J_{n+3} & J_{n+4} & J_{n+5} & J_{n+6} & J_{n+7} \\ J_{n+3} & J_{n+4} & J_{n+5} & J_{n+6} & J_{n+7} & J_{n+8} \\ J_{n+4} & J_{n+5} & J_{n+6} & J_{n+7} & J_{n+8} & J_{n+9} \\ J_{n+5} & J_{n+6} & J_{n+7} & J_{n+8} & J_{n+9} & J_{n+10} \end{vmatrix} = (-1)^{n+1} \quad (28)$$

Proof: Consider the determinant –

$$\Delta = \begin{vmatrix} 1 & 1 & 1 & 1 & 1 & 1 \\ 1 & 0 & 0 & 0 & 0 & 0 \\ 0 & 1 & 0 & 0 & 0 & 0 \\ 0 & 0 & 1 & 0 & 0 & 0 \\ 0 & 0 & 0 & 1 & 0 & 0 \\ 0 & 0 & 0 & 0 & 1 & 0 \end{vmatrix}$$

The value of this determinant is 1, we have

$$\Delta^2 = \begin{vmatrix} 2 & 2 & 2 & 2 & 2 & 2 \\ 1 & 1 & 1 & 1 & 1 & 1 \\ 1 & 0 & 0 & 0 & 0 & 0 \\ 0 & 1 & 0 & 0 & 0 & 0 \\ 0 & 0 & 1 & 0 & 0 & 0 \\ 0 & 0 & 0 & 1 & 0 & 0 \end{vmatrix}$$

Now, by mathematical induction,

$$\Delta^n = \begin{vmatrix} J_{n+1} & K_{n+1} & L_{n+1} & M_{n+1} & N_{n+1} & J_n \\ J_n & K_n & L_n & M_n & N_n & J_{n-1} \\ J_{n-1} & K_{n-1} & L_{n-1} & M_{n-1} & N_{n-1} & J_{n-2} \\ J_{n-2} & K_{n-2} & L_{n-2} & M_{n-2} & N_{n-2} & J_{n-3} \\ J_{n-3} & K_{n-3} & L_{n-3} & M_{n-3} & N_{n-3} & J_{n-4} \\ J_{n-4} & K_{n-4} & L_{n-4} & M_{n-4} & N_{n-4} & J_{n-5} \end{vmatrix}$$

Now, writing $M_{n+1} = J_n + J_{n-1}$ the R.H.S. can be written as the sum of two determinants, one of which is zero, Therefore,

DOI: http://dx.doi.org/10.5772/intechopen.110481

$$\Delta^n = \begin{vmatrix} J_{n+1} & K_{n+1} & L_{n+1} & M_{n+1} & J_{n-1} & J_n \\ J_n & K_n & L_n & M_n & J_{n-2} & J_{n-1} \\ J_{n-1} & K_{n-1} & L_{n-1} & M_{n-1} & J_{n-3} & J_{n-2} \\ J_{n-2} & K_{n-2} & L_{n-2} & M_{n-2} & J_{n-4} & J_{n-3} \\ J_{n-3} & K_{n-3} & L_{n-3} & M_{n-3} & J_{n-5} & J_{n-4} \\ J_{n-4} & K_{n-4} & L_{n-4} & M_{n-4} & J_{n-6} & J_{n-5} \end{vmatrix}$$

Now, writing $M_{n+1} = J_n + J_{n-1} + J_{n-2}$, the R.H.S. can be written as the sum of three determinants, two of which are zero. Therefore,

$$\Delta^n = \begin{vmatrix} J_{n+1} & K_{n+1} & L_{n+1} & J_{n-2} & J_{n-1} & J_n \\ J_n & K_n & L_n & J_{n-3} & J_{n-2} & J_{n-1} \\ J_{n-1} & K_{n-1} & L_{n-1} & J_{n-4} & J_{n-3} & J_{n-2} \\ J_{n-2} & K_{n-2} & L_{n-2} & J_{n-5} & J_{n-4} & J_{n-3} \\ J_{n-3} & K_{n-3} & L_{n-3} & J_{n-6} & J_{n-5} & J_{n-4} \\ J_{n-4} & K_{n-4} & L_{n-4} & J_{n-7} & J_{n-6} & J_{n-5} \end{vmatrix}$$

Now, writing $L_{n+1} = J_n + J_{n-1} + J_{n-2} + J_{n-3}$Ln + 1, the R.H.S. can be written as the sum of four determinants, three of which are zero. Therefore,

$$\Delta^n = \begin{vmatrix} J_{n+1} & K_{n+1} & J_{n-3} & J_{n-2} & J_{n-1} & J_n \\ J_n & K_n & J_{n-4} & J_{n-3} & J_{n-2} & J_{n-1} \\ J_{n-1} & K_{n-1} & J_{n-5} & J_{n-4} & J_{n-3} & J_{n-2} \\ J_{n-2} & K_{n-2} & J_{n-6} & J_{n-5} & J_{n-4} & J_{n-3} \\ J_{n-3} & K_{n-3} & J_{n-7} & J_{n-6} & J_{n-5} & J_{n-4} \\ J_{n-4} & K_{n-4} & J_{n-8} & J_{n-7} & J_{n-6} & J_{n-5} \end{vmatrix}$$

Now, writing $K_{n+1} = J_n + J_{n-1} + J_{n-2} + J_{n-3} + J_{n-4}$ the R.H.S. can be written as the sum of five determinants, four of which are zero. Therefore,

$$\Delta^n = \begin{vmatrix} J_{n+1} & J_{n-4} & J_{n-3} & J_{n-2} & J_{n-1} & J_n \\ J_n & J_{n-5} & J_{n-4} & J_{n-3} & J_{n-2} & J_{n-1} \\ J_{n-1} & J_{n-6} & J_{n-5} & J_{n-4} & J_{n-3} & J_{n-2} \\ J_{n-2} & J_{n-7} & J_{n-6} & J_{n-5} & J_{n-4} & J_{n-3} \\ J_{n-3} & J_{n-8} & J_{n-7} & J_{n-6} & J_{n-5} & J_{n-4} \\ J_{n-4} & J_{n-8} & J_{n-8} & J_{n-7} & J_{n-6} & J_{n-5} \end{vmatrix}$$

On arranging, we get

$$\Delta^n = \begin{vmatrix} J_{n+1} & J_n & J_{n-1} & J_{n-2} & J_{n-3} & J_{n-4} \\ J_n & J_{n-1} & J_{n-2} & J_{n-3} & J_{n-4} & J_{n-5} \\ J_{n-1} & J_{n-2} & J_{n-3} & J_{n-4} & J_{n-5} & J_{n-6} \\ J_{n-2} & J_{n-3} & J_{n-4} & J_{n-5} & J_{n-6} & J_{n-7} \\ J_{n-3} & J_{n-4} & J_{n-5} & J_{n-6} & J_{n-7} & J_{n-8} \\ J_{n-4} & J_{n-5} & J_{n-6} & J_{n-7} & J_{n-8} & J_{n-9} \end{vmatrix}$$

Putting, n-9 = m or n = m + 9 and substituting all the Δ's, we obtain,

$$(-1)^{m+9} = \begin{vmatrix} J_{m+10} & J_{m+9} & J_{m+8} & J_{m+7} & J_{m+6} & J_{m+5} \\ J_{m+9} & J_{m+8} & J_{m+7} & J_{m+6} & J_{m+5} & J_{m+4} \\ J_{m+8} & J_{m+7} & J_{m+6} & J_{m+5} & J_{m+4} & J_{m+3} \\ J_{m+7} & J_{m+6} & J_{m+5} & J_{m+4} & J_{m+3} & J_{m+2} \\ J_{m+6} & J_{m+5} & J_{m+4} & J_{m+3} & J_{m+2} & J_{m+1} \\ J_{m+5} & J_{m+4} & J_{m+3} & J_{m+2} & J_{m+1} & J_m \end{vmatrix}$$

Rearranging the determinant and replacing m with n we get the required result (28)

2.3 Generating matrix {Cn}

In this section, we will obtain some identities with the help of generating matrix, we consider the matrix,

$$[T] = \begin{vmatrix} 1 & 1 & 1 & 1 & 1 & 1 \\ 1 & 0 & 0 & 0 & 0 & 0 \\ 0 & 1 & 0 & 0 & 0 & 0 \\ 0 & 0 & 1 & 0 & 0 & 0 \\ 0 & 0 & 0 & 1 & 0 & 0 \\ 0 & 0 & 0 & 0 & 1 & 0 \end{vmatrix} \quad (29)$$

By mathematical induction, we can show that:

$$[T]^n = \begin{vmatrix} J_{n+1} & K_{n+1} & L_{n+1} & M_{n+1} & N_{n-1} & J_n \\ J_n & K_n & L_n & M_n & N_{n-2} & J_{n-1} \\ J_{n-1} & K_{n-1} & L_{n-1} & M_{n-1} & N_{n-3} & J_{n-2} \\ J_{n-2} & K_{n-2} & L_{n-2} & M_{n-2} & N_{n-4} & J_{n-3} \\ J_{n-3} & K_{n-3} & L_{n-3} & M_{n-3} & N_{n-5} & J_{n-4} \\ J_{n-4} & K_{n-4} & L_{n-4} & M_{n-4} & N_{n-6} & J_{n-5} \end{vmatrix} \quad (30)$$

where $n \geq 5$
and

$$[C_n, C_{n-1}, C_{n-2}, C_{n-3}, C_{n-4}, C_{n-5}] = [T]^n [C_5, C_4, C_3, C_2, C_1, C_0], n \geq 5 \tag{31}$$

On using (30) and (31), we get:

$$\begin{vmatrix} C_{n+P} \\ C_{n+P-1} \\ C_{n+P-2} \\ C_{n+P-3} \\ C_{n+P-4} \\ C_{n+P-5} \end{vmatrix} = \begin{vmatrix} J_{n+1} & K_{n+1} & L_{n+1} & M_{n+1} & N_{n-1} & J_n \\ J_n & K_n & L_n & M_n & N_{n-2} & J_{n-1} \\ J_{n-1} & K_{n-1} & L_{n-1} & M_{n-1} & N_{n-3} & J_{n-2} \\ J_{n-2} & K_{n-2} & L_{n-2} & M_{n-2} & N_{n-4} & J_{n-3} \\ J_{n-3} & K_{n-3} & L_{n-3} & M_{n-3} & N_{n-5} & J_{n-4} \\ J_{n-4} & K_{n-4} & L_{n-4} & M_{n-4} & N_{n-6} & J_{n-5} \end{vmatrix} \begin{vmatrix} C_n \\ C_{n-1} \\ C_{n-2} \\ C_{n-3} \\ C_{n-4} \\ C_{n-5} \end{vmatrix}$$

From this we obtain:

$$C_{n+P} = J_{P+1}D_n + K_{P+1}D_{n-1} + L_{P+1}D_{n-2} + M_{P+1}D_{n-3} + N_{P+1}D_{n-4} + J_{P+1}D_{n-5} \tag{32}$$

Let us now consider the matrix $[W]$, which is the transpose of the matrix [T] in,

$$[W] = [T] = \begin{vmatrix} 1 & 1 & 0 & 0 & 0 & 0 \\ 1 & 0 & 1 & 0 & 0 & 0 \\ 1 & 0 & 0 & 1 & 0 & 0 \\ 1 & 0 & 0 & 0 & 1 & 0 \\ 1 & 0 & 0 & 0 & 0 & 1 \\ 1 & 0 & 0 & 0 & 0 & 0 \end{vmatrix}$$

It can be shown that the sequence

$$C_4, P_5, Q_5, , R_5, S_5, C_5 \ldots ., C_{n-1}, P_n, Q_n, , R_n, S_n, C_n \tag{33}$$

is generated by matrix $[W]$

$$[C_n, P_n, Q_n, R_n, S_n, C_{n-1}] = [W]^{n-5} [C_5, P_5, Q_5, R_5, S_5, C_4], n \geq 5 \tag{34}$$

On using (33) and (34), we get

$$[C_{n+P}, P_{n+P}, Q_{n+P}, R_{n+P}, S_{n+P}, C_{n+P}]$$
$$= [W]^P [C_n, P_n, Q_n, R_n, S_n, C_{n-1}]$$
$$= \begin{vmatrix} J_{P+1} & J_P & J_{P-1} & J_{P-2} & J_{P-3} & J_{P-4} \\ K_{P+1} & K_P & K_{P-1} & K_{P-2} & K_{P-3} & K_{P-4} \\ L_{P+1} & L_P & L_{P-1} & L_{P-2} & L_{P-3} & L_{P-4} \\ M_{P+1} & M_P & M_{P-1} & M_{P-2} & M_{P-3} & M_{P-4} \\ N_{P+1} & N_P & N_{P-1} & N_{P-2} & N_{P-3} & N_{P-4} \\ J_P & J_{P-1} & J_{P-2} & J_{P-3} & J_{P-4} & J_{P-5} \end{vmatrix} \begin{vmatrix} C_n \\ P_n \\ Q_n \\ R_n \\ S_n \\ C_{n-1} \end{vmatrix}$$

$$C_{n+P} = J_{P+1}C_n + J_P P_n + J_{P-1}Q_n + J_{P-2}R_n + J_{P-3}S_n + J_{P-4}C_{n-1}$$
$$P_{n+P} = K_{P+1}C_n + K_P P_n + K_{P-1}Q_n + K_{P-2}R_n + K_{P-3}S_n + K_{P-4}C_{n-1}$$
$$Q_{n+P} = L_{P+1}C_n + L_P P_n + L_{P-1}Q_n + L_{P-2}R_n + L_{P-3}S_n + L_{P-4}C_{n-1}$$
$$R_{n+P} = M_{P+1}C_n + M_P P_n + M_{P-1}Q_n + M_{P-2}R_n + M_{P-3}S_n + M_{P-4}C_{n-1}$$
$$S_{n+P} = N_{P+1}C_n + N_P P_n + N_{P-1}Q_n + N_{P-2}R_n + N_{P-3}S_n + N_{P-4}C_{n-1}$$

Application:
We can introduce generalized Fibonacci n-step polynomials. Based on generalized Fibonacci n-step polynomials, we can define a new class of square matrix of order n and we can state a new coding theory called generalized Fibonacci n-step theory.

3. Discussion

Mathematics has enormous potential for solving the various problems of daily life. The Fibonacci polynomials are a polynomial sequence that can be considered as generalization sequences worked upon by many mathematicians earlier like as Atanassov [11], Harne & Parihar [4], and Georgiev and Atanassov [8] in accordance with our findings. The chapter has wider acceptance for the fruitful study of various case studies as illustrated in the current citation, which is well supported by the earlier studies too.

4. Conclusions

There are many known identities for the Fibonacci recursion relation. We define the sequence $\{C_n\}$ and its four comparison sequences $\{P_n\}$, $\{Q_n\}$, $\{R_n\}$, and $\{S_n\}$. We drive linear sum properties of comparison sequence. We also derive generating matrix for the sequence $\{C_n\}$.

Author details

Manjeet Singh Teeth[1*] and Sanjay Harne[2]

1 Department of Mathematics, Christian Eminent College Affiliated to Devi Ahilya University, Indore, India

2 Department of Mathematics, Mata Jijabai Government P.G. Girls College, Affiliated to Devi Ahilya University, Indore, India

*Address all correspondence to: manjeetsinghteeth@gmail.com

References

[1] Vorobyov NN. The Fibonacci Numbers. Boston Pergamon: D.C. Health Company; 1963

[2] Waddill ME, Lovis S. Another generalized fibonacci sequence. The Fibonacci Quarterly. 1967;**5**(3):209-222

[3] Jaiswal DV. On a generalized fibonacci sequence, Labdev. Journal of Science and Technology. 1969;**7**(2):67-71

[4] Harne S, Parihar CL. Generalized fibonacci sequence. Ganit Sandesh (India). 1994;**8**(2):75-80

[5] Teeth MS, Harne S. Polynomial related to generalized fibonacci sequence. Journal of Ultra Scientist of Physical Sciences. 2022;**34**(2):16-24

[6] Georghiou C. On some second order linear recurrence. The Fibonacci Quarterly. 1989;**27**(2):10-15

[7] Georgiev P, Atanassov KT. On one generalization of the Fibonacci sequence, Part III. Some relations with fixed initial values. Bulletin of Number Theory and Related Topics. 1992;**16**:83-92

[8] Georgiev P, Atanassov KT. On one generalization of the Fibonacci sequence, Part II. Some relations with arbitrary initial values. Bulletin of Number Theory and Related Topics. 1995;**16**:75-82

[9] Georgiev P, Atanassov KT. On one generalization of the Fibonacci sequence. Part V. Some examples. Notes on Number Theory and Discrete Mathematics. 1996a;**2**(4):8-13

[10] Georgiev P, Atanassov KT. On one generalization of the Fibonacci sequence, Part VI: Some other examples. Notes on Number Theory and Discrete Mathematics. 1996b;**2**(4):14-17

[11] Atanassov KT. An arithmetic function and some of its applications. Bulletin of Number Theory and Related Topics. 1985;**9**(1):18-27

[12] Atanassov KT, Atanassov L, Sasselov D. A new perspective to the generalization of the Fibonacci sequence. The Fibonacci Quarterly. 1985;**23**(1): 21-28

[13] Walton JE, Horadam AF. Some aspects of generalized fibonacci numbers. The Fibonacci Quarterly. 1974a;**12**(3):241-250

[14] Walton JE, Horadam AF. Some further identities for the generalized Fibonacci sequence $\{H_n\}$. The Fibonacci Quarterly. 1974b;**12**(3):272-280

Chapter 8

Information Encoding for Flow Watermarking and Binding Keys to Biometric Data

Boris Assanovich, Iryna Korlyukova and Andrei Khombak

Abstract

Due to the current level of telecommunications development, fifth-generation (5G) communication systems are expected to provide higher data rates, lower latency, and improved scalability. To ensure the security and reliability of data traffic generated from wireless sources, 5G networks must be designed to support security protocols and reliable communication applications. The operations of coding and processing of information during the transmission of both binary and non-binary data in nonstandard communication channels are described. A subclass of linear binary codes is considered, which are both Varshamov-Tenengolz codes and are used for channels with insertions and deletions of symbols. The use of these codes is compared with Hidden Markov Model (HMM)-based systems for detecting intrusions in networks using flow watermarking, which provide high true positive rate in both cases. The principles of using Bose-Chadhuri-Hocquenhgem (BCH) codes, non-binary Reed-Solomon codes, and turbo codes, as well as concatenated code structures to ensure noise immunity when reproducing information in Helper-Data Systems are considered. Examples of biometric systems organization based on the use of these codes, operating on the basis of the Fuzzy Commitment Scheme (FCS) and providing FRR < 1% for authentication, are given.

Keywords: linear codes, Varshamov-Tenengolz codes, non-binary turbo codes, Reed-Solomon codes, concatenated codes, flow watermarking, biometric system

1. Introduction

Engineers and researchers around the world have been using various error correction codes (ECCs) for almost a century to provide communication and combat noise in information channels. In addition to communication, ECCs have found many other uses, including watermarking and intrusion detection, cryptography, and information security. Digital watermarking is the process of embedding a digital code into some public data. Today, this technology is widely used not only in multimedia processing but also in network traffic monitoring. In this case, the input patterns, which are easily identified when the watermarked flows cross an observation point, allow the creation of a mechanism to scan the network for the harmful activity. This procedure finds applications both for securing network connections and intrusion detection in them.

On the other hand, when providing secure access to any data, it becomes necessary to use user verification by analyzing his password, which requires ensuring the reliability of its storage. To solve this problem, biometric methods of organizing secure access to the system are widely used, that reduce the risks of storing passwords, which have long been a weak point in security systems. This chapter will discuss some types of the ECC and how they can be used to help ensure the security and reliability of information.

In recent years, the technique of applying the ECC has been undergoing changes due to the use of machine learning (ML) methods and, in particular, deep learning (DL). A good review of the recent advancements in DL-based communication was made by Qin et al. [1], where the authors described the use of this technique for channel modeling, modulation recognition, and improvement of decoding methods. In recent papers, the authors have considered in more detail the DL methods for decoding known codes [2] and, moreover, for constructing an ECC based on intelligent methods [3]. Despite the increasing use of the ML technique for ECC, it is important to understand both the principles of describing known ECC based on algebraic constructions that lead to elegant decoding algorithms and their application in non-standard communication channels.

The rest of the chapter is organized as follows. First, we present the basic encoding-decoding principles of the binary and non-binary ECC used for substitution and symbol insertions and deletions errors in Section 2. Then we discuss the flow watermarking techniques for intrusion detection in Section 3. In Section 4, we describe the use of various ECC types in biometric systems (BSs) for solving the problem of authentication and present our conclusion in Section 5.

2. Error-correcting codes

2.1 Linear codes

At the present stage of the ECC theory and technology development, more and more complex code structures attract our attention. Although coding algorithms are becoming more complex and require powerful computing resources, in recent years, researches have increasingly turned to known codes and mathematical descriptions developed for them. Such codes, for example, are *linear codes*, which have useful properties and can be used in non-standard data transmission channels applications.

There are many good tutorials about error-correcting codes (for example, see [4, 5]), so only the necessary definitions are used in the entire chapter. We define a *code* C of block length n over an alphabet q is a subset of q^n, together with a one-to-one encoding which maps a message set M to a code set C. The main goal of encoding is to increase the resilience of the messages to errors, where $|C|$ denotes the number of elements in a set or the code cardinality.

We start from the description of linear code. A linear q-ary code of length n and dimension k is a linear subspace C with dimension k of the vector space with dimension n, whose elements are the elements of the field $GF(q)$ The description of the properties of linear codes will be done on the example of binary codes, whose symbols are the elements of a field $GF(2) = \{0;1\}$ which is a code alphabet.

Generally, a binary code C is defined as a set of finite sequences (vectors) $\mathbf{x} = (x_1, \dots, x_n)$, called codewords, encoded with the use of corresponding message vectors $\mathbf{b} = (b_1, \dots, b_k)$ from code symbols x_i, $b_i \in GF(2)$. Linear (n,k,d)-code is defined by following parameters: Hamming distance between binary codewords $d(\mathbf{x}_i;\mathbf{x}_j)$, weight of a codeword $wt(\mathbf{x}_i)$ and a code rate or coding efficiency k/n. Linear codes are defined by their generator and parity-check matrices **G** and **H**, respectively, whose columns and rows are linearly independent. Every codeword is a linear combination of rows of the generator matrix **G**. The *minimum distance* $d_{min} = \min\{d(\mathbf{x}_i;\mathbf{x}_j)\}$ of a linear ECC and its code weight distribution define its error correction capacity t or maximum number of symbols that can be corrected in a codeword. There is a simple method of minimum distance decoding with syndrome that could be applied in order to correct t or less errors in a codeword.

According to this principle, the decoder selects a codeword to minimize the Hamming distance of the matched codeword relative to the received codeword **y** using a reduced look-up table. This is allowed by the linear property of the code.

The decoder performs following steps: the syndrome calculation of codeword **y**:

$$\mathbf{S} = \mathbf{y} \cdot \mathbf{H}^{\mathrm{T}}, \tag{1}$$

determination of the most likely error vector **e**, and estimation of the possibly transmitted codeword $\mathbf{x}^*$. Next, the decoder selects that vector **e** of the smallest weight that satisfies $\mathbf{e} \cdot \mathbf{H}^{\mathrm{T}} = \mathbf{S}$. These syndrome-based decoding procedures are linear and of low complexity, and only the second step requires a non-linear look-up table operation. In the case of linear codes use, the so-called standard arrays are widely applied, which make it possible to find the corresponding codeword for any received vector.

The standard array for a binary (n,k) code is an array of size 2^{n-k} by 2^k where: (1) the first row has the codewords with "all zeros" on the left); (2) the 1st column is a coset leader for a coset in each row; and (3) the entry in the i-th row and the j-th column is the sum of the i-th adjacency coset leader and the j-th codeword. However, the linear property of the code allows the use of syndrome decoding, which is an efficient decoding technique using a reduced look-up table.

For linear codes, it is important that the number of syndromes, 2^{n-k}, must be greater than or equal (for perfect codes) to the number of correctable error patterns $\sum_{i=0}^{t} \binom{n}{i} \leq 2^{n-k}$, which is determined by the so-called Hamming bound [4].

If we take a linear (6,3,3)-code C with codewords {(000000), (110100), (011010), (101110), (101001), (011101), (110011), and (000111)}, obtained on the basis of the generator matrix **G** ([5], pp. 357–367), then there are modification methods to change its properties [5]. For example, the number of its codewords can be increased or decreased. If individual codewords are removed from code set C, then a new code C' with the same properties can be constructed while maintaining the minimum weight of codewords. This modified code C' is a *subcode* of C.

2.2 Cyclic codes

Binary *cyclic codes* are block codes for which cyclic shifts of each codeword yield a different codeword and can be efficiently encoded and decoded using shift registers and combinatorial logic. Cyclic codes are linear codes with good properties and can be defined by polynomials:

$$u(x) = u_0 + u_1x + \ldots + u_{n-1}x^{n-1} \tag{2}$$

In such a polynomial representation, the presence or absence of the formal variable x with a degree is determined by the coefficient and corresponds to the binary “1” or “0” of the codeword element.

Cyclic codes have the property that all code polynomials $u(x)$ are multiples of a unique polynomial $g(x)$, called the *generator polynomial* of the code. This generator polynomial is completely described by its roots, which are called zeros of the code.

Sometimes, to find a generating polynomial $g(x)$, the polynomial $(x^n - 1)$ must be factored into its irreducible factors $f_i(x)$. Since a cyclic code is also linear, any set of linearly independent vectors can be selected as a generator matrix. However, in this case, a nonsystematic encoding is performed, when the message bits can appear explicitly in any positions of a codeword. However, the encoding of codewords of a binary cyclic code can be also systematic, if the message is processed in another way. With this encoding, information and check symbols are clearly separated. Another polynomial, $h(x)$, called the parity-check polynomial, can be related to the parity-check matrix. Generator polynomial and parity-check polynomial are connected by $g(x)h(x) = x^n + 1$.

Then, a parity-check matrix for a cyclic code is given by using as rows the binary vectors associated with the first $n - k - 1$ nonzero cyclic shifts. In the case of high-rate cyclic (n, k) codes, say $k/n > 0.5$, encoding by the division of $x^{n-k}u(x)$ by $g(x)$ or by recursion with $h(x)$, the coefficients of $u(x)$ are in the systematic form so that the first k coefficients are the message bits and the remaining $n - k$ coefficients are the control bits. However, for powerful cyclic codes correcting multiple errors, the algebraic decoding procedure becomes much more complicated.

It should be noted that the principles of representation and encoding and decoding of polynomial codes are based on the concepts of both simple and extended finite fields, calculations in which can be found in a number of textbooks [5]. Below, we will only briefly use the basic concepts.

Representatives of more powerful correction codes are the Bose-Chadhuri-Hocquenhgem (BCH) codes that provide suitable selection of block lengths, code rates, and correcting capacity. BCH codes are cyclic codes that are constructed by specifying the roots of their generator polynomials, i.e., a BCH code of $d_{min} \geq 2t_d + 1$ is a cyclic code whose generator polynomial $g(x)$ has $2t_d$ consecutive roots $\alpha^b, \alpha^{b+1}, \alpha^{b+2t_d-1}$, where t_d is a designed capacity. Next, the generator polynomial of the BCH (n, k, d_{min}) code is

$$g(x) = LCM\left\{f_b(x), f_{b+1}(x), \ldots, f_{b+2t_d-1}(x)\right\}. \quad (3)$$

Here, LCM is the least common multiple. Thus, we have a code with a length of $n = LCM\{n_b, n_{b+1}, \ldots, n_{b+2t_d-1}\}$, and dimension of $k = n - \deg[g(x)]$ and a designed minimum distance $2t_d + 1$, which in the general case can be less than the real minimum distance.

For example, consider GF(2^4), $p(x) = x^4 + x + 1$, with $t_d = 2$ and $b = 1$. Then,

$$g(x) = LCM\{(x^4 + x + 1)(x^4 + x^3 + x^2 + x + 1)\} = x^8 + x^7 + x^6 + x^4 + x + 1. \quad (4)$$

We get a double-error-correcting binary BCH (15,7,5) code.

The main idea of decoding binary BCH codes is to use the elements of $GF(2^m)$ to number the positions of a codeword that are found by solving a set of equations, which can be obtained from the error polynomial and the zeros of the code. The most popular methods for decoding BCH codes include the Berlekamp-Massey (BM), Euclid, and Peterson-Gorenstein-Zierler (PGZ) algorithms and are discussed in more detail in [4].

2.3 Reed-Solomon codes

Reed-Solomon codes are multiple error-correcting non-binary codes that were introduced by Irving S. Reed and Gustave Solomon in 1960. There are two main representations of Reed-Solomon codes – the original representation and the BCH-based representation, which is the most common, due to the fact that BCH-based decoding is more efficient compared to the original representation decoders. In the first case, if $u(x) = u_0 + u_1x + \ldots + u_{k-1}x^{k-1}$ is given as an information polynomial, and $u_i \in GF(2^m)$, then there are 2^{mk} such polynomials obtained after calculating $u(x)$ over nonzero elements of $GF(2^m)$, which are codewords of the RS$(2^m - 1, k, d)$ code of length 2^m. If we interpret RS codes as non-binary BCH codes and the values of code coefficients are taken from $GF(2^m)$, then zeros for a t_d error-correcting RS code are $2t_d$ consecutive powers of α. Moreover, since over $GF(2^m)$ the minimal polynomials have the form $f_i(x) = (x - \alpha^i),\ 0 \leq i < 2^m - 1$ and for some integer b, which usually have values of 0 or 1, we have [4]

$$g(x) = \prod_{j=b}^{b+2t_d-1} (x + \alpha^j), \tag{5}$$

It follows from Eq. (3) that the minimum distance of RS (n, k, d) code over GF(2^m) is $d \geq n - k + 1$. On the other hand, RS code satisfies singleton bound [4] with equality $d = n - k + 1$, which defines it as a maximum distance separable (MDS) code. Since the Reed-Solomon code is a linear code, it is possible to apply the classical coding procedure using its generator matrix.

The decoding algorithms of RS codes are similar to that of binary BCH codes. As shown above, setting the primitive powers of the root as evaluation points makes the Reed-Solomon source code cyclic. Reed-Solomon codes in BCH representation are always cyclic because BCH codes are cyclic. In this regard, they are characterized by the same decoding methods as for cyclic codes. In order to choose the correct algorithm that meets the requirements of the system, it is necessary to understand its purpose, which is determined by the RS decoder operation. There are cycle decoding evaluation algorithm, PGZ algorithm, BM algorithm, Sugiyama algorithm with erasures and without erasures, and list decoding algorithms.

Reed Solomon code can correct not only errors but also the erasures, i.e., so-called "lost" symbols. If n_{er} symbols of RS code are erased and the remaining $n - n_{er}$ symbols contain n_e errors, the BM algorithm can find the correct codeword as long as $n_{er} + 2n_e \leq 2t < d$. If $n_{er} = 0$, the decoder is used as an errors-only decoder, and if $0 < n_{er} \leq d - 1$ we can call the decoder as an error-and-erasure decoder (EED) [6].

Sudan in 1997 introduced an algorithm that allows the correction of errors beyond the minimum distance of the code. This algorithm produces a list of codewords (it is a list decoding algorithm) and is based on interpolation and factorization of polynomials over $GF(2^m)$ and its extensions. The main idea of such decoding is to create a list of possible codewords and apply a list-decoding algorithm with such characteristic as a (ρ, L)-list, where ρ is a fractional value of the Hamming distance and L is the size of the list. It was shown [7] that if the fraction of errors in the received information is at most ρ, then the transmitted codeword is guaranteed to be in the output list. Also, note that if C is (ρ,L)-list decodable, then we can output at most L codewords for any received codeword by Sudan algorithm. Application of this algorithm allows to correct $n - 2\sqrt{nk}$ errors. Several years later, Guswami and Sudan improved the algorithm to correct up to $n - \sqrt{nk}$ errors [7].

The algebraic decoding methods described above are generally hard decision decoding (HDD) methods, which means that for each symbol a hard decision is made about its value. However, the decoder may also contain an information about the reliability of symbol (for example, the demodulator's confidence in the correctness of the symbol), which allows to build soft decision decoders (SDDs). The advent of turbo codes that use iterated soft decision propagation decoding techniques to achieve error correction efficiency has spurred interest in applying SDD to conventional algebraic codes.

2.4 Turbo codes

Turbo codes involve the concatenation of two recursive systematic convolutional (RSC) codes connected serially or in parallel, and an interleaver between them. Due to space limitations in this section, we omit the description of convolutional codes. The iterative decoding of constituent codes starts individually, either serially or in parallel, based on inputs derived from the channel and typically some a priori information. Information from each data symbol propagates through the overall code structure in time. The optimal decoding algorithm for each component code in terms of minimizing the probability of error given independent inputs is the Bahl-Cocke-Jelinek-Raviv (BCJR) algorithm [8], realizing the maximum a posteriori (MAP) criterion decoding. Then resulting symbol probabilities are used to find the log-likelihood ratio (LLR) for q-1 candidate values when decoding them. Next [8], the most likely element is determined by comparing each LLR value and selecting the symbol with the highest reliability (highest LLR).

2.5 VT codes

Often, to describe and compare codes, a channel model is used in which information is transmitted. However, in the presence of noise in the channel, symbols may be received with errors. This type of error sometimes called the substitution error. The influence of interference in communication channels also causes synchronization errors associated with the insertion of additional symbols or deletion of transmitted symbols, which are sometimes called "indels." Therefore, there is a strong reason to develop codes that not only correct substitution errors but also deal with "indels."

One of the first codes to deal with synchronization errors caused by symbol deletion was the Varshamov-Tenengol'ts (VT) codes. Below, we briefly consider this construction.

Given a parameter a, with $0 \leq a \leq n$, the Varshamov-Tenegol'ts (VT) code $\mathrm{VT}_a(n)$ is the set of binary words $\mathbf{x} = (x_1, \ldots, x_n)$ of length n so that the equality satisfies [9]:

$$\sum_{i=1}^{n} ix_i \equiv a(\mathrm{mod}(n+1)). \tag{6}$$

These codes are single-error-correcting codes and optimal for $a = 0$ as it was conjectured in [10] and will be discussed below.

For example, after calculation $\sum_{i=1}^{n} ix_i \equiv 0(\text{mod}7)$ with length n = 6, we can get VT code set $\text{VT}_0(6)$ = {(000000), (001100), (010010), (011110), (100001), (101101), (110011), (110100), (111111)}. Any code $\text{VT}_0(n)$ can be used to communicate reliably over a channel that introduces at most one "indel" in a block of length n. Levenshtein proposed a simple decoding algorithm [11] based on the deficiency in checksum and weight calculation for a VT code. As an example, assume the code $\text{VT}_0(6)$ is used and $\mathbf{x} = (110100) \in \text{VT}_0(6)$ is transmitted over the channel. If the first bit in **x** is deleted and **y** = (10100) is received, then the new checksum is 4, and the deficiency D = 7–4 = 3 > $wt(\mathbf{y})$ = 2. The decoder must insert a binary "1" after n-D = 3 "0's" from the right to get a codeword (110100). Such an algorithm for decoding $\text{VT}_0(n)$ code with deletion correction is based on a shift operation and has low complexity.

Considering the simplicity of calculating the parameters of VT codes, we would like to make a linear encoder for efficient mapping of binary message sequences into codewords. For binary VT codes, such an encoder was proposed by Abdel-Ghaffar and Ferriera [12]. They constructed a systematic encoder that maps k-bit message sequences onto codewords in $\text{VT}_a(n)$, for $k = n - \lceil \log_2(n+1) \rceil$, where in parentheses is rounding up to a higher integer. In addition, for these codes, the concept of a *syndrome* can be used, which is found as $Syn(C) \equiv \sum_{i=1}^{n} ix_i \mod(n+1)$.

Now we can introduce the "parity" bits denoted by $t_p = n - k = \lceil \log_2(n+1) \rceil$ and use then in dyadic positions to ensure that Syn(C) = a. Therefore, the message bits can be encoded by calculating the value of the difference between the desired syndrome and calculated one $d_C = a - Syn(C) \mod(n+1)$.

In an example, see [12] of code for n = 10 and a = 0, the parity check and information positions can be represented, respectively, as {1, 2, 4, 8} and {3, 5, 6, 7, 9, 10}, and used to encode **b** = (011001) as follows: x = ($x_1x_20x_4110x_801$), where.

$x_1 + 2x_2 + 4x_4 + 8x_8 = 0 - (3 \cdot 0 + 5 \cdot 1 + 6 \cdot 1 + 7 \cdot 0 + 9 \cdot 0 + 10 \cdot 1 = 1(\text{mod}11)$.

The parity-check sequence of least lexicographic order ($x_1x_2x_4x_8$) = 011000 can be taken.

However, $\text{VT}_0(n)$ codes are nonlinear, and the dimension of k for obtaining linear (n,k) codes is limited as $k \leq \lfloor n/2 \rfloor$ [13]. Below, we propose an algorithm for finding a linear *substitution and deletion/insertion correction code* from any existed $\text{VT}_0(n)$. The proposed algorithm is executed step by step as follows: 1) sort the codewords of the code $\text{VT}_0(n)$ in lexicographic order; 2) find and choose k linearly independent codewords of maximum weight while maintaining $d(\mathbf{x}_\mathrm{i}; \mathbf{x}_\mathrm{j}) \geq d_{min}$; and 3) construct matrices **G** and **H** from C, making linear combinations of the selected VT codewords.

Using this algorithm will allow constructing a subcode that has at least k + 1 codewords of the $\text{VT}_0(n)$ code. Obviously, the linear combination of any codeword with itself forms a codeword (0 ... 0), which is also belongs to the code $\text{VT}_0(n)$. By exploiting the algorithm proposed above, the following generator and parity-check matrixes for the modified (6,3,3)-code C' have been constructed:

$$\mathbf{G}' = \begin{bmatrix} 1 & 1 & 0 & 1 & 0 & 0 \\ 1 & 0 & 1 & 0 & 1 & 0 \\ 0 & 1 & 1 & 0 & 0 & 1 \end{bmatrix}, \mathbf{H}' = \begin{bmatrix} 1 & 0 & 0 & 1 & 1 & 0 \\ 0 & 1 & 0 & 1 & 0 & 1 \\ 0 & 0 & 1 & 0 & 1 & 1 \end{bmatrix}. \quad (7)$$

Representing **G**′ and **H**′ as Eq. (5) results in a code set with an increased number of codewords belonging to $\text{VT}_0(6)$ compared to initial code C. If we discard the elements that are not $\text{VT}_0(6)$ codewords and the codeword (000000), and then we get a subcode consisting of four codewords with desired properties: {(110100), (011110), (101101), (110011)}.

Thus, C^* is a linear subcode with $d_{min} = 3$, at the same time it is a $VT_0(6)$ code. Therefore, it can be used to correct one substitution error and one "indel" error. At the same time, the analysis showed that its code rate is reduced by about ½ compared to the code rate of C. The proposed algorithm [14] can be applied to an arbitrary code to find a correcting VT code, which is a subcode of a linear code. If we take a linear ECC (8,2,5) [5 , p.378], consisting of four codewords, we can also find a linear subcode for it, which is also the code $VT_0(8)$. Its properties of one "indel" error and two substitution errors correction are preserved. It is known that the size of any $VT_0(n)$ is about $2^n/n$ [6], then additional properties appear, decreasing its rate to less than ½.

Recently, these codes have again attracted interest, as evidenced by the publication [15], where an encoding method was proposed for a non-binary systematic VT code.

3. Use of error-correcting coding in flow watermarking

3.1 HMM-based model for watermark embedding and extraction

The watermark embedding algorithm aims to detect any changes in the marked data or its integrity. The contents integrity is performed in the verification process. In this section, we discuss the application of ECC for watermark embedding in the context of traffic analysis (TA) used for such purposes as diagnostic monitoring, resource management, and intrusion detection. Intrusion detection systems attempts to detect intrusion through analyzing the network traffic with the use of watermark tracing [16]. If the embedded watermark is both reliable and unique, it is possible to analyze the watermarked return traffic and trace it back at intermediate nodes. This TA approach is referred to as the "flow watermarking" (FW).

To prevent an attacker to endure and analyze the delayed packets and then to eliminate the embedded watermarks, the developed FW schemes have to be "invisible" in the network. An example of stepping-stone detection scenario with FW is depicted in **Figure 1** where an Attacker attacks Victim hiding his identity. Fortunately, FW can be applied for tracing back the attack source.

FW is often implemented on the basis of *inter-packet-delay* (IPD) schemes [17], where watermark bits are embedded in the intermediate packet time which allows to hide traffic artifacts from an attacker. However, in this case, the replacement of

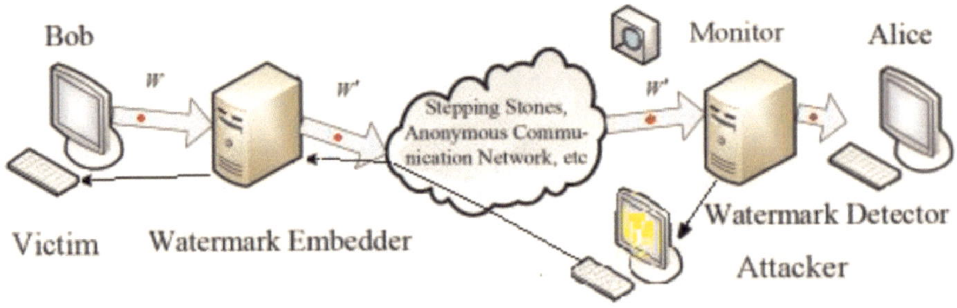

Figure 1.
Attacker detection scenario.

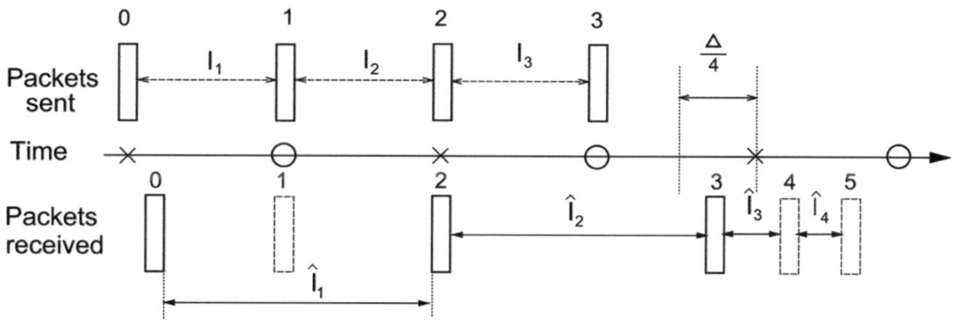

Figure 2.
An example of IPDs distortion.

packets and packet loss can cause severe detection and decoding errors. The use of ECC makes it possible to improve the noise immunity of FW systems.

The presence of contiguous packet merging leads to a telecommunication channel with deletion and/or substitution errors, and the appearance of jitter-induced bursting or splitting of packets also causes symbol insertions, which requires the appropriate choice of coding for reliable transmission of watermarks. **Figure 2** demonstrates these phenomena. It follows from it that four packets 0, 1, 2, and 3 are sent, three packets 0, 2, and 3 are received, packet 1 is lost, and new packets 4 and 5 are added.

Most FW technologies use a carrier that modulates the transfer of watermark data. Gong et al. [18] embedded quantization index modulation (QIM) watermarks into IPDs and added a layer of ECC to handle watermark desynchronization and substitution errors. Authors developed a Hidden Markov Model (HMM) for channel with dependent deletion and substitution errors using a maximum likelihood decoding (MLD) algorithm paired with a forward-backward algorithm for the calculation of the posterior probabilities [5]. The schematic of the proposed system can be depicted as shown in **Figure 3**.

This scheme uses an Encoder and Decoder to process the incoming watermark sequences in order to obtain the codewords $\mathbf{x}$ of the applied ECC. With this notification, it is implied that the length of a watermark w and a codeword $\mathbf{x}$ is defined by user and corresponds to some selected value. Further in the text, superscripts are used to indicate the length of the sequence, and subscripts to determine the element number in a sequence.

Due to the network artifacts described above, the additional transformations must be performed in the encoder to improve noise immunity. For example, see [19], a *spasification* procedure based on one-hot coding is implemented so that a sparse version of w is formed, denoted as w^N, where $N = sn$ and s is the *sparsification factor* and has an integer value. This procedure for a channel with the presence of insertions, deletions, and substitutions (IDS) can also been extended for the non-binary case [20].

The s value is controlled by the density f that is the ratio of "ones" in w^N:

$$f = \sum_{i=1}^{N} w_i^N / N. \tag{8}$$

The whole scheme uses a secret key k, known to the Encoder and Decoder, which is added bit by bit to w^N forming a binary sequence x^N containing one or more codewords.

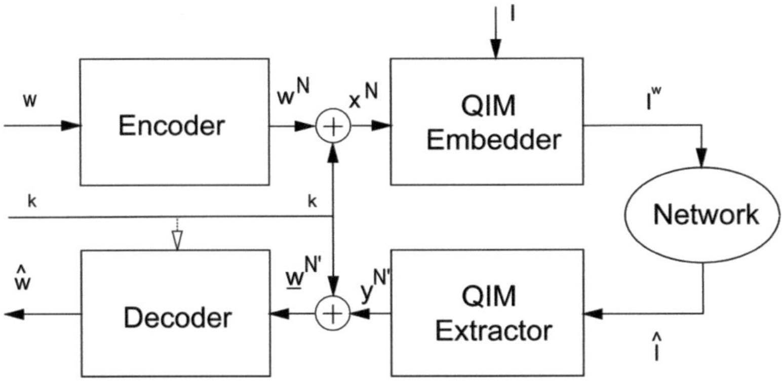

Figure 3.
System diagram.

The key used for security plays a supporting role in dealing with IDS channel errors during decoding. For example, see [18], if w_1 = '1', c = 8, then x^N = 10000000 and if k = 11111011 then x^8 = 01111011. When one bit is dropped and y^7 = 0111011 is received, it can be supposed with a high probability that bit "1" from the second series of bits was lost.

Next, we will consider in more detail the principle of QIM for which modulation and demodulation are carried out by means of QIM Embedder and QIM Extractor, respectively (see **Figure 3**).

To embed the watermark, the IPD flow is modified so that each IPD is converted to an interval according to the even/odd multiplier of the quantization interval Δ/2, depending on the value of the 0/1 bit. Formally, this can be represented as:

$$I_i^w = \begin{cases} c\Delta, & \text{if} \quad x_i = 0 \\ (c + 0.5)\Delta, & \text{if} \quad x_i = 1 \end{cases} \tag{9}$$

Since packets can only be delayed by the QIM Embedder, it is possible to define the c parameter to be the smallest integer so that the change in I_i^w would slightly delay the i-th packet. After passing the I^w sequence through the network, it is received as an estimated IPD sequence $\hat{I}$ and then analyzed by the QIM extractor, obtaining the necessary information from it.

The following QIM demodulation threshold function is used to recover the embedded bit y_i:

$$y_i = \begin{cases} \mathrm{mod}\left(\lfloor 2\hat{I}_i/\Delta \rfloor, 2\right) & \text{if} \quad 2\hat{I}_i/\Delta - \lfloor 2\hat{I}_i/\Delta \rfloor \leq 0.5 \\ \mathrm{mod}\left(\lceil 2\hat{I}_i/\Delta \rceil, 2\right) & \text{if} \quad 2\hat{I}_i/\Delta - \lfloor 2\hat{I}_i/\Delta \rfloor > 0.5 \end{cases} \tag{10}$$

Consider the example in **Figure 2**. Here, the first two IPDs I_1 and I_2 are converted into $\hat{I}_1$, and the size of the last IPD I_2 is changed and is determined as $\hat{I}_2$. Therefore, the result of the noise in the channel is the bit received before Packet 2, which is due to the two intervals merging $y_1 = x_1 \oplus x_2$, and the bit inversion after receiving Packet 3, resulting in $y_2 = x_3$.

In general $y_i = \sum_{j=r+1}^{i} x_j$ and can take the binary values "0" or "1", where r is the index of the last successfully received packet before the i-th one. As can be seen from

Figure 2, two intervals $\widehat{I}_3, \widehat{I}_4$ appear resulting in the insertion of new bits into the received watermarked data.

Obviously, in the absence of packet loss or split, the watermark bit is inverted if the IPD jitter exceeds Δ/4. The jitter can be described by i.i.d. Laplace distributed with zero mean. Then the jitter substitution error probability can be estimated as:

$$P_s = 1 - F(\Delta/4) = 0.5\exp\left(-\Delta/2\sqrt{2}\sigma\right), \tag{11}$$

where $F()$ is the Laplacian pdf and σ^2 is its variance. Since packet losses leads to merging of successive IPD, the resulting error contains both deletion and substitution error [18]. In this model, we assume that packets are lost independently and that the initial packet is always synchronized.

Authors in [17, 18] used the concept of *drift* to define the loss of bit synchronization, which is the shift in position of some sent packet in the received flow. Using sparse key parameters, one can determine the probabilities of IDS events in the resulting sequence. These events were interpreted with the use of HMM and applying the forward-backward algorithm [19], the watermark estimation posterior probabilities for the maximum likelihood decoding (MLD) have been derived as.

$$\widehat{w}_j = \arg \quad \max P\left(y^{N'}|w_j\right), \quad w_j \in \{0, 1\}. \tag{12}$$

After calculating these probabilities for all bits of the watermark sequence, the presence of a watermark in flow is determined based on the correlation value of the resulting sequence and the original one. For those interested in the details of mathematical calculations, one can refer to the original publications of the authors mentioned above.

3.2 Use of VT codes in FW

An alternative IPD-FW scheme for embedding watermarks based on the use of binary VT codes, which are subcodes of linear codes and exploiting QIM, has been proposed in [14]. The scheme uses linear codes of length 6 and 8 bits with an attached marker and optional matrix interleaving to deal with bursting errors.

Coding-decoding scheme without interleaving. As before, we assume that the watermark w to be embedded is a bit sequence. Next, the sequence w is divided into blocks of bits $\mathbf{b} = (b_1 \dots b_l)$ of length l and encoded by the chosen VT code $\mathbf{x}$ of length n (see above). Then obtained codeword $\mathbf{x}$ is concatenated with predefined marker pattern z of length m making w^N, where $N = n + m$. In this implementation, a pattern z contains a series of zeros, which is determined by the necessity of XOR-ing all bits of the formed sequence w^N with a secret key k, by analogy with the previous HMM-based method. The key k used is a sparse sequence containing a binary "1" in only one position out of all N bits. In fact, the sequence w^N can be made up by the concatenation of M codewords $\mathbf{x}$ with a marker z attached. We denote this composite sequence as $w^N = w_1 w_2 \dots w_M$, with which the composite secret key is XOR-ed, forming the sequence $x^N = x_1 x_2 \dots x_M$.

Next, the generated sequence x^N enters the QIM Embedder, where modulation is performed in the same way as described above. Then the IPD sequence I^w with injected watermark pattern is transmitted and after traversing the network is received in the form of estimated sequence $\hat{I}$ and demodulated. The result sequence $y^{N'}$ is xored with a key sequence k and serves as an input to the Decoder.

The Decoder detects markers in the $w^{N'}$ sequence and separates it into codewords of the applied VT code. As a result, the codewords derived from $w^{N'}$ can contain substitutions, insertions, and deletions. The sequence at the Encoder output x^N as well as the one that enters the Decoder $y^{N'}$ in general case do not match. In addition, their lengths may differ, which makes the decoding process difficult.

To solve the problem, it is proposed to use hybrid decoding with error correction and the choice of one of two algorithms is depended on the number of errors in each received codeword $\mathbf{y}$ [14]. The decoder-type selection is based on an estimate of the codeword $\mathbf{y}$ length. If the only one "indel" is found, the Levenshtein's decoding algorithm [11] is used, and if the number of "indel" errors is greater than 1, the MLD $\mathbf{x}^* = \arg\max \quad \Pr(\mathbf{x}/\mathbf{y})$ is applied. The syndrome decoding (see Eq. (1)) is performed in case of the absence of "indel" errors or after they have been corrected.

For example, suppose that a sequence $w^{N'}$ = 110100000.11110000 at the output of the QIM Extractor processed with the key k = 000000000.001000000 to be decoded using the subcode mapping C' = {(110100), (110011), (011110), (101101)} into message blocks $\mathbf{b}$ = {00, 01,10,11}. After detecting a marker and removing it, two codewords $\mathbf{y}_1$ = 110100, $\mathbf{y}_2$ = 11010 are obtained. The syndrome calculation (Eq. (1)) of $S = 0$ can serve as a flag that the boundaries of the received word $\mathbf{y}_1$ are not changed, there are no errors in it, or the number of errors exceeds its corrective capacity. Therefore, it is possible to apply the Levenshtein decoding algorithm to correct the deleted bit in the last position of $\mathbf{y}_2$. However, if one more bit is also deleted, after estimating the length $\mathbf{y}_2$, it is necessary to proceed to use MLD decoding.

Coding-decoding scheme with interleaving. Considering the channel with bursts of errors, the effective mechanisms for separating error bursts are the use of interleaving. We consider an approach using matrix interleaving of a linear subcode of the VT code, which simplifies the decoding process.

It was found in [21] that there is a VT code that coincides with a linear (8,2,5) error-correcting code [5], consisting of four codewords and subcoding $\mathrm{VT}_0(8)$. However, to perform the independent decoding of codewords from a linear VT subcode, placed in a continuous bitstream, the boundaries of the codewords must be known. We can implement their independent decoding by the organization of the codewords set of linear subcode and the use of matrix interleaving.

The proposed scheme consists of several layers. However, to simplify its work, we describe it based on the scheme in **Figure 3**. As before, we assume that a watermark sequence w^N is divided into segments of messages $b = b_1 \ldots b_l$ and encoded by the VT Encoder forming codewords $\mathbf{x}$ of length n. These codewords are written row by row into an $Q \times n$ interleaving matrix. Next, each column is concatenated with a predefined marker pattern, which increases the number of matrix rows. Then matrix columns are XOR-ed with the fragments of the secret key k represented as a sparse binary sequence with a small number of binary ones. The resulting version of w^N is then read column by column from the interleaving matrix forming a sequence x^N.

In fact, x^N is a supercode containing Q codewords, which is embedded in flow IPD via QIM Embedder. Note that the elements of interleaving and deinterleaving are not shown in the scheme of **Figure 3**. Further, after processing in QIM Embedder, passing through the network, the IPD flow is demodulated in the QIM Extractor and undergoes inverse transformations with respect to encoding (XORing with key, marker removal, finding codeword boundaries, deinterleaving, and decoding). For

P_d (synthetic traffic)	1%	2%	3%	10%	20%
HMM-based	1.000	1.000	1.000	0.994	—
VT code	0.999	0.999	0.999	0.995	0.666

Table 1.
TPR values for varying P_d *with FPR* < *1%.*

information, various interleaving schemes and adjacent deletions correction constructions have been discussed in [22].

Two FW methods have been modeled: the first one is based on HMM and the second one uses VT codes with markers. At the same time, in the first method, the length of the sparse sequence for FW was 10 bits, and in the second method, it was 9 bits, considering the $VT_0(6)$ code with 3-bit marker. About 5000 packets were generated, in which network jitters were modeled as Laplace distribution with zero mean and a standard deviation of 10 ms. In the synthetic channel, substitution errors followed sequentially after deletion errors, and symbol insertions were studied separately. The detection threshold was chosen to keep false positive rate (FPR) below 1% for all deletion probabilities. The evaluation of true positive rates (TPRs) in the detection of watermarks for two schemes with respect to different deletion probabilities P_d is presented in **Table 1**.

It follows from **Table 1** that the use of less complex VT coding leads to virtually the same performance compared to HMM. From the results, the TPR value drops to 66% when the packet loss is 20%, which is rare in a network environment. Methods using interleaving and code (8,2,5) showed better results [21] for channels with bursting insertion errors.

4. Application of error-correcting codes in biometrics

In recent years, there has been increasing interest in cryptographic approaches using biometric measurements. For these purposes, many physical methods are used: from taking fingerprints of a person to the dynamics of his gait. The uniqueness of these characteristics allows them to be used for both identification and authentication. However, for the verification organization, it is required to perform the recognition procedure. A special *biometric template*, which is a mathematical representation of features from the original data, is used to store biometric characteristics.

In this section, we will focus on the processing of biometric features of a person's face. Face recognition is very flexible and can be performed from a distance. These systems can be classified as follows [23]: image-based matching (whole face), feature-based face recognition, and video-based matching. The accuracy of the user's biometric data recognition is high. However, the security and privacy of user data may be compromised. In this case, the concept of *cancelable biometrics* is applicable.

The idea of a reversible template was proposed by Ratha et al. [24]. It includes five main features: tautology, irreversibility, accuracy, diversity, and revocability.

There are several approaches to the creation of biometric system (BS), which are based on direct generation of a secret key from biometrics or key binding to biometric data. The widespread implementation of BS solutions is constrained by the fuzziness of biometric data. This problem can be alleviated by applying error correction codes. Below we will consider several BS based on the use of different methods for obtaining biometric features and various code structures using the so-called *Fuzzy Commitment Scheme* (FCS) [25].

4.1 Biometric system based on facial HOG features

The use of ECC is due to the spread of biometric measurement values, which can be regarded as noise added to the received signal. Taking into account the signal processing procedures for registration and verification, the generalized scheme can be represented as shown in **Figure 4**.

Let us consider the operation of the BS in accordance with [26] with the only difference that instead of local information from convolutions with Gabor kernels, the histogram of oriented gradients (HOG) is used as features. In addition, more powerful BCH codes are used to suppress noise due to fuzzy biometric data [27].

The principle of the scheme operation is as follows. The Preprocessor receives the set of images of the user's face as input, scales them, and converts color images into gray scale ones. Next, HOG features are extracted from the images in the form of real Y sequences, which can be represented as vectors with a dimension of 4464 elements.

The Preprocessor calculates mean $\vec{\mu}_i$ and variance $\vec{s}_i$ for the series of biometric data samples submitted by each i-th user, as well as the global mean $\vec{\mu}$ for all registered users.

In addition, the reliability function R_i is calculated here for each bit p of each user in according to the expression:

$$R_{i,p} = \frac{1}{2}\left(1 + erf\left(\left\{(\vec{\mu}_i)_p - (\vec{\mu})_p\right)\right\}/\sqrt{2s_{i,p}^2}\right)\right). \quad (13)$$

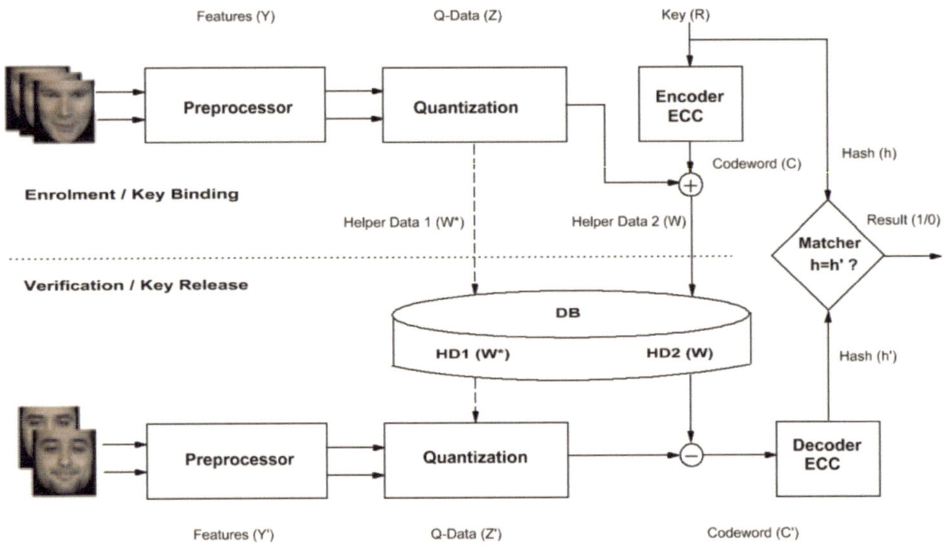

Figure 4.
Diagram of a biometric system.

Based on the calculated parameters (Eq. (13)), according to Y_j values, a *mask* is formed containing reliable numbers of data positions in Y based on a selected threshold. As a result, the obtained values of $\vec{\mu}_i$, $\vec{\mu}$, and mask information form Helper data 1 (W^*) that are written to the database (DB) before the Quantization procedure. The Quantizer performs data binarization according to Eq. (14) forming Z sequence of length n:

$$(Q_i)_p = 0, \quad if \quad (\vec{\mu}_i)_p \leq (\vec{\mu})_p, \quad and \quad (Q_i)_p = 1, \quad if \quad (\vec{\mu})_p > (\vec{\mu})_p \tag{14}$$

At the same time, a codeword C and its hash value $h = h(C)$ are formed by using the ECC Encoder from the user secret key R. Then a Helper data 2 (W) is calculated using the XOR operation as follows $W_i = C_i \oplus Z_i$. As a result, the enrollment procedure is completed and the values W^*, W, $h(C)$ are entered into the database.

When implementing the user verification, one or more images are sent to the Preprocessor, where they are converted into a sequence of real numbers Y'. Based on W^*, reliable positions are determined, Q_j data values are binarized, and $Z'$is obtained. Next, the values of HD2 are retrieved from the database, and operations $C'_i = W_i \oplus Z'_i$ are performed. The codeword C' is decoded by the ECC Decoder, and its hash value $h' = h(C')$ is calculated. Verification is considered successful if $h(C)$ and $h(C')$ matches and the corresponding user key R is extracted.

The OpenFace tool [28] was used to obtain the HOG characteristics of user images containing 12×12 blocks of 31 histograms and written into a row vector Y of length 4464 real values. BCH codes (511,58) and (511,28) over $GF(2^m)$ were applied as ECC, correcting $t_d = 91$ and $t_d = 118$ errors. For performance testing, the Caltech database was used with face images of 24 users. Inter-class and intra-class distributions of the fractional Hamming distance were obtained, which, together with the verticals of the applied BCH codes, are shown in **Figure 5**.

The calculated values of false acceptance rate (FAR) and false rejection rate (FRR) had the following values: FRR = 0, FAR = 3.5% demonstrating good performance of the used BCH codes, allowing to choose the lengths of secret keys $K_1 = 58$ and $K_2 = 28$ bits. Obviously, the length of the K_2 key is too small to register a large number of users. It is clear that after binary quantization the real data are highly rounded, which leads to significant quantization noise. To adapt to biometric real features, we further used unquantized real data processing and non-binary turbo encoding.

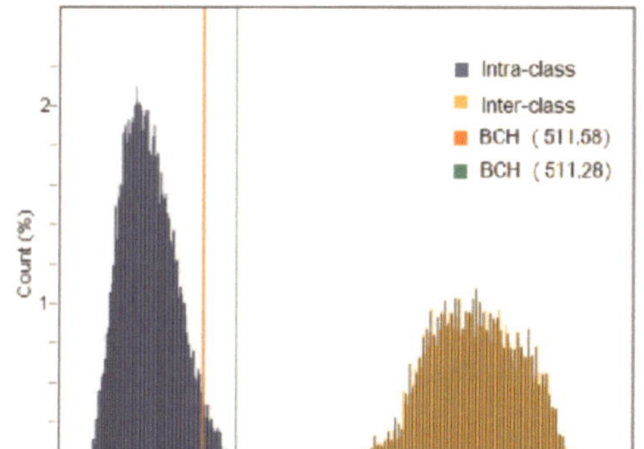

Figure 5.
Inter-class and intra-class distributions.

4.2 Application of turbo codes in biometric systems

Recently, the use of non-binary turbo codes with modulation in a biometric system has been studied [29]. These codes were generated from non-binary convolutional component codes combined with a random interleaver. Then phase-shift keying (8-PSK) was used. During processing, the polynomials with a coding matrix $g = [166;176]$ over the ring $GF(8)$ were used, which made it possible to implement a systematic turbo code with rate of 1/3. As a result, a random octal secret key of length 166 was encoded in turbo code with trailing zeros, forming the resulting 3x172 matrix at the output, that was then modulated into the 8-PSK constellation [30]. Each symbol of turbo code X was presented by I-Q complex numbers giving framed data matrix 3x344. To get biometrical face features, the Caltech database has been used. Data from 511 real numbers obtained after masking procedure (see above) to get components of 4464-element HOG vectors have been used as biometric raw data B. Then the quantized data with the interval $q = 0.19635$ was normalized and linearly mapped to the interval [0, 2pi) of angles presented then by 2 I-Q components. Hence, the hashed value of result code block together with quantized real data is put into public DB.

At the authentication stage, the resulted codeword Y corrupted by "biometric noise" B' is iteratively decoded by the modified BCJR algorithm giving the user password and a hash value. The main operations on 8-ary data blocks (vectors) according to the principles of the BS scheme are shown in **Figure 6**.

Preliminary experimental estimates of FRR resulted in value FRR $\sim$ 0.1%, which is several times better than the previous scheme and known results for turbo codes [31].

A further increase in the effectiveness of BS is possible by increasing the inter-class differences in biometric characteristics, which prompted the use of neural networks (NNs) in this area.

In the NN-based system below, we have applied the stacked autoencoder (SAE) structure and the concatenated ECC using RS codes.

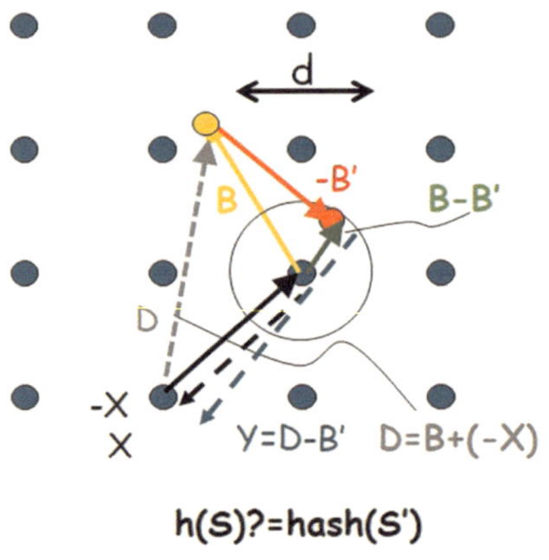

Figure 6.
Vector processing of modulated real data for turbo codes.

4.3 Smiling face biometric authentication system

In the following BS [6], we consider the use of a stacked autoencoder (SAE) to extract features from a sequence of video frames of a user smiling face in order to authenticate him and provide the access to digital services. In contrast to the generally accepted application of binary ECC in BS, we used the non-binary Reed-Solomon codes concatenated with the binary linear ones. The use of these codes, taking into account the dimension of the symbols, leads to an increase in the length of their bit representation. On the other hand, in order to neutralize "biometric noise" and correct errors, it is necessary to increase the ECC redundancy, which reduces the encoding efficiency.

The biometric templates are created according to FCS [25] based on the equidistant quantization of real data at the output of SAE for further processing and encoding by concatenated RS codes. The BS uses concatenated ECC based on non-binary RS codes and binary linear codes with the use of hard decision decoding (HDD) technique and soft decision decoding (SDD) obtained from symbol reliability. The operation principle the proposed BS will be described based on the generalized authentication scheme shown in **Figure 4**.

The Preprocessor block performs such operations as video data capturing, face and smile detection, the smile frames extraction, image transformation and normalization, and features extraction using the SAE pretrained at the registration stage. The biometric data samples obtained from the SAE output layer form the concatenated supervector Y = $\{Y^1, \dots, Y^M\}$ from several vectors Y^i, where M is a number of processed frames. At the stage of Quantization, the real data Y are converted into their quantized versions Z producing also deviations W^* of data values relative to their mean values or centers of quantization intervals used as HD1. In the Encoder block, the user's password or Key R is encoded with one or more ECC codewords, depending on the required password strength and FRR. Further, for a biometric authentication purposes, the bit representation of the resulting codeword is XOR-ed with the encoded version C of quantized data Z, which results in W that serves as HD2.

At the Enrollment stage, the biometric samples obtained at the SAE output using HD1 and HD2 then are binded to the secret Key R. In addition, the h-hash value of the codewords is calculated and stored in the biometric database DB. During verification, the reverse process of decoupling the "auxiliary" data HD1, HD2, decoding codewords C', h'-hash calculation, and comparison of two hashes h and h' are performed.

A series of experiments were carried out with SAE to get good compact biometric features. To reduce time spent, in these experiments, the subsets of 40 subjects were randomly selected from the entire UvA-NEMO database [32], reproducing a user smile. Unsupervised learning results and then supervised tuning of SAE with parameters 127/63 in the form of histograms are shown in **Figure 7** showing the significant expansion of the inter-class distributions relative to each other.

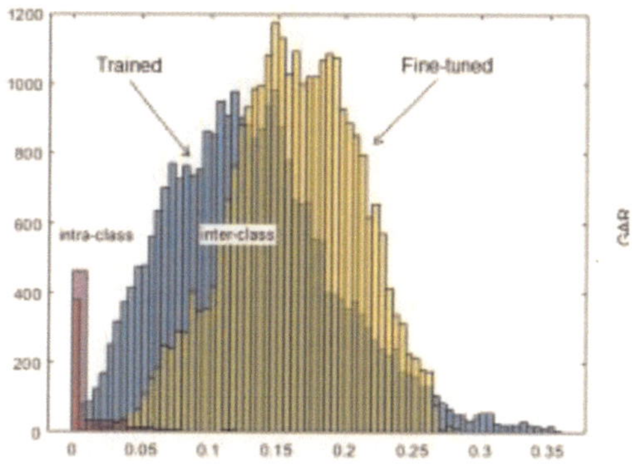

Figure 7.
Shift of inter-class distributions during SAE training.

The expected values of FAR and FRR were estimated based on the block error probability of decoding for the uncorrectable error patterns accepted by BS. In the evaluation, we conducted simulation experiments for different error-correcting code structures. The results are placed in **Table 2** [6].

From **Table 2**, it follows that reducing the RS code length makes it possible to increase the performance of the BS in terms of the FRR parameter. Simulation experiments have shown the possibility of achieving the FRR less than 1% for key lengths of 90–170 bits and demonstrated a more efficient use of RS codes compared to the previous scheme and the results from [33] for face template protection.

Inner code	Outer code	FRR, %	Key, bit/frames × dimension	Efficiency
RS (63,15)	Linear (6,3,1)	1.0	90/2 × 63(180/4 × 63)	0.119
RS (63,15)	REP (3,1,1)	0.5	90/3 × 63(180/3 × 63)	0.0079
RS (31,9)	REP (3,1,1)	0.5	90/6 × 31	0.0968
RS (63,21)	—	0.7	126/1 × 63	0.33
RS (31,17)	—	0.3	170/2 × 31	0.5
RS (31,17)	REP (3,1,1)	<0.1	170/6 × 31	0.1828

Table 2.
Evaluation of FRR and key size for different ECC structures.

For all studied schemes, privacy leakage was assessed. The calculated mutual information between the input (output) data was significantly less than the entropy of the ECC codewords, which actually confirms the impossibility of compromising the user biometric data.

5. Conclusion

Despite the fact that the development of error-correcting codes was aimed at application in communication systems, their use is also relevant in security systems, where it is required to neutralize the noise added to the data from the environment. In this chapter, the main code structures that have found application in the flow watermarking for network intrusion detection, as well as in biometric authentication systems, have been considered.

The watermarking environment model is treated as a channel with substitution, insertion, and deletion errors. Two main code constructions were considered, first: based on HMM with adding a synchronizing key sequence to sparse data and second: based on the use of the modified error-correcting VT codes with a marker attachment. Statistical and computational experiments have shown the same performance of these schemes in terms of watermark detection TPR ≈ 1 when FPR $< 1\%$ with a simpler implementation of the second scheme, which is slightly inferior in coding rate to the first one. At the same time, the considered implementations of FW schemes are invisible and sufficiently resistant to network artifacts if their relative values do not exceed 20%.

In addition, two types of face biometric authentication systems based on HOG structures and latent autoencoder data were considered. The fuzziness of the HOG data was compensated by using binary BCH codes (511,58) and (511,28), which made it possible to obtain the FRR parameter value of 3.5%. The use of non-binary turbo codes of rate 1/3 with octal data modulation provided the possibility to improve performance up to the value of FRR = 1% with real helper data. And the use of concatenated RS codes together with linear binary codes showed the possibility of increasing efficiency and achieving FRR values of less than 1%. Moreover, it has been shown that a decrease in the FRR parameter is possible, firstly, by increasing the redundancy of the concatenated ECC, and secondly, by using the additional information from helper data when exploiting the EED for RS codes.

Thus, the transition to efficient non-binary code structures and real-valued ECC is a promising area of research in the field of watermarking and biometrics.

Acknowledgements

The authors express their gratitude to the publisher for financial support.

Conflict of interest

The authors declare no conflict of interest.

Author details

Boris Assanovich*, Iryna Korlyukova and Andrei Khombak
Yanka Kupala State University of Grodno, Grodno, Belarus

*Address all correspondence to: bas@grsu.by

References

[1] Qin Z, Ye H, Li GY, Juang B-HF. Deep learning in physical layer communications. IEEE Wireless Communications. 2019;**26**(2):93-99. DOI: 10.1109/MWC.2019.180060

[2] Nachmani E, Marciano E, Lugosch L, Gross WJ, Burshtein D, Y. Be'ery. Deep learning methods for improved decoding of linear codes. IEEE Journal of Selected Topics in Signal Processing. 2018;**12**(1): 119-131. DOI: 10.1109/ JSTSP.2017.2788405

[3] Huang L, Zhang H, Li R, Ge Y, Wang J. AI coding: Learning to construct error correction codes. IEEE Transactions on Communications. 2020;**68**(1):26-39. DOI: 10.1109/TCOMM.2019.2951403

[4] Morelos-Zaragoza RH. The Art of Error Correcting Coding. 2nd ed. Chichester, West Sussex, England: Wiley; 2006. p. 278. DOI: 10.1002/ 0470035706

[5] Sklar B. Digital Communications: Fundamentals and Applications. 2nd ed. Upper Saddle River, N.J.: Prentice-Hall PTR; 2001. p. 1079

[6] Assanovich B, Kosarava K. Authentication System Based on Biometric Data of Smiling Face from Stacked Autoencoder and Concatenated Reed-Solomon Codes. PRIP'2021. CCIS, 1562. In: Proceedings of the 15th International Conference. Cham: Springer; 2021. pp. 205-219

[7] Guruswami V, Rudra AM. Sudan Essential Coding Theory. University at Buffalo; 15 Mar 2019. p. 473. eBook (Creative Commons Licensed, 2022)

[8] Carrasco RA, Johnston M. Non-Binary Error Control Coding for Wireless Communication and Data Storage. Chichester, West Sussex, United Kingdom: Wiley; 2008. p. 322

[9] Varshamov RP, G.M. Tenengol'ts. Correction code for single asymmetric errors. Avtomat. Telemekh. 1965;**26**(2): 286-290

[10] G. M. Tenengol'ts. Class of codes correcting bit loss and errors in the preceding bit. Avtomat. Telemekh. 1976; **37**(5):797-802

[11] Levenshtein VI. Binary codes capable of correcting deletions, insertions and reversals. Soviet Physics-Doklady. 1966; **10**(8):707-710

[12] Abdel-Ghaffar KAS, Ferreira HC. Systematic encoding of the Varshamov-Tenengolts codes and the Constantin-Rao codes. IEEE Transactions on Information Theory. 1998;**44**:340-345

[13] Abdel-Ghaffar KAS, Ferreira HC, Cheng L. Correcting deletions using linear and cyclic codes. IEEE Transaction on Information Theory. 2010;**56**(10): 5223-5234

[14] Assanovich B, Puech W, Tkachenko I. Use of linear error-correcting subcodes in flow watermarking for channels with substitution and deletion errors. In: Proceedings 14th IFIP TC 6/TC 11 Int. Conf. Commun. Multimedia Security (CMS). Magdeburg, Germany; 2013. pp. 105-112

[15] Abroshan M, Venkataramanan R, Fabregas AGI. Efficient systematic encoding of non-binary VT Codes. In: 2018 IEEE International Symposium on Information Theory (ISIT). Vail, CO, USA: IEEE; 17-22 Jun 2018. pp. 91-95. DOI: 10.1109/ISIT.2018.8437715

[16] Wang X, Reeves DS, Wu SF, Yuill J, Sleepy watermark tracing: An active network-based intrusion response framework. In: Proceedings Trusted Information New Decade Challenge IFIP TC11 16th Annu. Working Conference Information. Paris, France: Security (IFIP/SEC); 2001. pp. 369-384

[17] Gong X, Rodrigues M, Kiyavash N. Invisible flow watermarks for channels with dependent substitution and deletion errors. In: Proceedings of the International Conference on Acoustics, Speech, and Signal Proc. Kyoto, Japan; 2012. pp. 1773-1776

[18] Gong X, Rodrigues M, Kiyavash N. Invisible flow watermarks for channels with dependent substitution, deletion, and Bursty insertion errors. IEEE Transactions on Information Forensics and Security. 2013;**8**(11): 1850-1859

[19] Davey MC, Mackay DJC. Reliable communication over channels with insertions, deletions, and substitutions. IEEE Transactions on Information Theory. 2001;**47**(2):687-698

[20] Yazdani R, Ardakani M. Reliable communication over non-binary insertion/deletion channels. IEEE Transactions on Communications. 2012; **60**(12):3597-3608

[21] Assanovich B, Terre VA, Penaranda-Foix FL. Watermarking pattern recognition in channels with substitution and bursty insertion and deletion errors. In: Proceedings of Pattern Recognition and Information Processing. Minsk; 2016. pp. 185-189

[22] Cheng L, Swart TG, Ferreira HC, Abdel-Ghaffar KA, Codes for correcting three or more adjacent deletions or insertions. In: IEEE International Symposium on Information Theory (ISIT). Honolulu, USA; Jul. 2014. pp. 1246-1250

[23] Tran QN, Turnbull BP, Hu J. Biometrics and privacy-preservation: How do they evolve? IEEE Open Journal of the Computer Society. 2021;**2**:179-191. DOI: 10.1109/OJCS.2021.3068385

[24] Ratha NK, Connell JH, Bolle RM. Enhancing security and privacy in biometrics-based authentication systems. IBM Systems Journal. 2001;**40**: 614-634

[25] Juels A, Wattenberg M. A fuzzy commitment scheme. In: Proceedings 6th ACM Conference Computer and Communications Security. Singapore; 2-4 Nov 1999. pp. 28-36. DOI: 10.1145/ 319709.319714

[26] Kevenaar TAM, Schrijen GJ, van der Veen M, Akkermans AHM, Zuo F. Face recognition with renewable and privacy preserving binary templates. In: Fourth IEEE Workshop on Automatic Identification Advanced Technologies (AutoID'05). Buffalo, NY, USA; 2005. pp. 21-26. DOI: 10.1109/ AUTOID.2005.24

[27] Assanovich B, Veretilo Y. Biometric database based on HOG structures and BCH codes. In: Proceedings of Information Technology and Systems (ITS2017). Minsk; 2017. pp. 286-287

[28] Baltrušaitis T, Robinson P, Morency L-P, OpenFace: An open source facial behavior analysis toolkit. In: 2016 IEEE Winter Conference on Applications of Computer Vision (WACV). Lake Placid, NY, USA; 2016. pp. 1-10. DOI: 10.1109/ WACV.2016.7477553

[29] Assanovich B, Veretilo Y. Fuzzy secure sketch biometric scheme based on non-binary turbo codes. In: Proceedings

of Information Technology and Systems. (ITS2018). Minsk; 2018. pp. 186-187

[30] Assanovich B. Application of Turbo codes for data transmission in UWB using PSK modulated complex wavelets, In. Signal Processing Workshop (SPW). 2020;**2020**:40-43. DOI: 10.23919/SPW49079.2020.9259136

[31] Maiorana E, Blasi D, Campisi P. Biometric template protection using Turbo codes and modulation constellations. IEEE WIFS. 2012;**2012**: 25-30

[32] Dibeklioğlu H, Salah AA, Gevers T. Recognition of genuine smiles. IEEE Transactions on Multimedia. 2015;**17**(3): 279-294

[33] Chen L et al. Face template protection using deep LDPC codes learning. IET Biometrics. 2019;**8**:190-197